Springer Textbooks in Earth Sciences, Geography and Environment

The Springer Textbooks series publishes a broad portfolio of textbooks on Earth Sciences, Geography and Environmental Science. Springer textbooks provide comprehensive introductions as well as in-depth knowledge for advanced studies. A clear, reader-friendly layout and features such as end-of-chapter summaries, work examples, exercises, and glossaries help the reader to access the subject. Springer textbooks are essential for students, researchers and applied scientists.

More information about this series at http://www.springer.com/series/15201

Charles Fox

Data Science for Transport

A Self-Study Guide with Computer Exercises

Charles Fox
Institute for Transport Studies
University of Leeds
Leeds
UK

ISSN 2510-1307 ISSN 2510-1315 (electronic)
Springer Textbooks in Earth Sciences, Geography and Environment
ISBN 978-3-030-10291-3 ISBN 978-3-319-72953-4 (eBook)
https://doi.org/10.1007/978-3-319-72953-4

© Springer International Publishing AG 2018
Softcover re-print of the Hardcover 1st edition 2018
This work is subject to copyright. All rights are reserved by the Publisher, whether the whole or part of the material is concerned, specifically the rights of translation, reprinting, reuse of illustrations, recitation, broadcasting, reproduction on microfilms or in any other physical way, and transmission or information storage and retrieval, electronic adaptation, computer software, or by similar or dissimilar methodology now known or hereafter developed.
The use of general descriptive names, registered names, trademarks, service marks, etc. in this publication does not imply, even in the absence of a specific statement, that such names are exempt from the relevant protective laws and regulations and therefore free for general use.
The publisher, the authors and the editors are safe to assume that the advice and information in this book are believed to be true and accurate at the date of publication. Neither the publisher nor the authors or the editors give a warranty, express or implied, with respect to the material contained herein or for any errors or omissions that may have been made. The publisher remains neutral with regard to jurisdictional claims in published maps and institutional affiliations.

Cover image: Simon and Simon photography
thestudio@simonandsimonphoto.co.uk
www.simonandsimonphoto.co.uk

Printed on acid-free paper

This Springer imprint is published by Springer Nature
The registered company is Springer International Publishing AG
The registered company address is: Gewerbestrasse 11, 6330 Cham, Switzerland

Foreword

There has never been a greater need to understand what is going on our networks whether it be highways, public transport, train, or other sustainable modes. Road users demand a greater level of information about the roads they are using and the services that they are using, and there is an expectation that they will be able to access this information real time and in a mobile environment. This creates significant challenges for local highway authorities in particular as they have traditionally not been early technology adopters, for good reasons, as they need to be sure that taxpayers' money is used appropriately and is not wasted.

Derbyshire County Council have identified the importance, opportunities, and advantages that transport data in its many forms can provide as well as helping to provide the basis for effective decision making and preparing for a "smarter" future. A key area of activity is Highways Asset Management, where timely information is being used to provide evidence to support the effective management of the highway asset, the highest value asset that the County Council operates. An example is the use of near real-time information to assess traffic conditions, while thinking how we can share this with the road user to proactively manage the network, which in turn may provide the opportunity to generate more capacity in our highway networks. A key aspect in being able to achieve this is a fundamental understanding of the data at our disposal, its advantages and limitations, and how we can most effectively process, manipulate, and communicate the information.

Information is key to all of our activities and can provide exciting opportunities to make considerable savings and efficiencies if appropriate techniques are adopted, the appropriate investment is made in technology, and very importantly, the right skills are developed to make sense of all the streams of information at our disposal. This book provides the foundations to get to grips with a myriad of data in its many forms, and the opportunities for greater insights and collaborations to be developed when bringing diverse data sets together. These are fundamental skills in the burgeoning field of data science, if we are to deliver on the "Big Data, Internet of Things, Smart Cities" agenda that is a current focus in the transport arena.

Derbyshire County Council, UK
Neill Bennett
Senior Project Officer
Transportation Data and Analysis

Preface

This book is intended for professionals working in Transportation who wish to add Data Science skills to their work, and for current or potential future students of graduate or advanced undergraduate Transport Studies building skills to work in the profession. It is based closely on a module of the same name which forms a core part of the MSc in Mathematical Transport Modeling course at the Institute for Transport Studies, University of Leeds, which trains many of the world's leading transport professionals. The live module was designed with the help of many leading transport consultancies in response to their real-world skills shortages in the area. It is taught to small groups of students in a highly interactive, discursive way and is focused around a single team project which applies all the tools presented to deliver a real system component for Derbyshire Council each year. Each chapter of this book presents theory followed computer exercises running on self-contained companion software, based on aspects of these projects.

The companion software provides a preinstalled transport data science stack including database, spatial (GIS) data, machine learning, Bayesian, and big data tools. No programming knowledge is assumed, and a basic overview of the Python language is contained within this book. This book is neither a complete guide to programming nor a technical manual for the tools described. Instead, it presents enough understanding and starting points to give the reader confidence to look up details in online software reference manuals or to participate in Internet communities such as *stackoverflow.com* to find details for themselves. This book is also intended to provide an overview of the field to transport managers who do not want to program themselves but may need to manage or partner with programmers. To enable this, computer examples are usually split into a separate section at the end of each chapter, which may be omitted by non-programmers.

This book follows roughly the structure of a typical Transport Data Science consulting engagement. Data Science consulting is a rapidly evolving field but has recently begun to stabilize into a reasonably standard high-level process. This begins with the data scientist finding the right questions to ask, including consideration of ethical issues around these questions. Often, a client does not know exactly what they need or want, or what is technically possible. At this stage, the data scientist must both gain an understanding of the business case requirements and convey to the client what is possible. This includes a professional responsibility to represent the field accurately and free from the hype which currently plagues many popular presentations. Next, suitable data must be obtained. This may be existing data owned by the client or available from outside, or in some cases will be collected in newly commissioned studies. Often, the data were collected for some other purpose and need work to make it usable for a new purpose. Sometimes, it will be in a format such as a series of Web pages which are intended for human rather than automated reading. Again, ethical issues around privacy and data ownership are important here, as well as technical aspects of linking together data from different sources, formats, and systems. Typically, this will involve "cleaning" of data, both to fix problems with individual sources such as missing or incorrect data and to ensure compatibility between data from different sources.

Some of the topics we will cover are:

- The relevance and limitations of data-centric analysis applied to transport problems, compared with other types of modeling and forecasting,
- Reusing and reprocessing transport data for Data Science,
- Ontological issues in classical database design,
- Statistics and machine learning analytics,
- Spatial data and GISs for transport,
- Visualization and transport map making,
- "Big data" computation,
- Non-classical "NoSQL" database ontologies,
- Professional and ethical issues of big data in transport.

For those following the programming details, some of the computational tools covered are:

- *Postgres*: How to design and create a relational database to house "raw" data,
- *SQL*: How to interrogate such data and provide appropriate data for transport modeling,
- *Python*: Basic programming skills and interfaces to Data Science tools,
- *PostGIS* and *GeoPandas*: spatial data extensions for SQL and Python for transport data,
- *scikit.learn*, *GPy*, and *PyMC3*: machine learning libraries,
- *Hadoop* and *Spark*: Big data systems.

The term "big data" is often used by the media as a synonym for Data Science. Both terms lack widely agreed formal definitions, but we consider them to be distinct here as follows.

"Data Science" as distinct from just "science" emphasizes the reuse of existing data for new purposes. Specifically, regular "science" is based on *causal* inference, in which some of the data are *caused* to have certain values by the scientist. This enables causal inferences to be made as a result: Causation is put into the system and can therefore be read out of the system. In contrast, Data Science typically makes use of passively observed data, which does not enable the same form of causal reasoning to take place.

"Big data" in this book means data which require the use of parallel computation to process. The actual numerical size (e.g., measured in bytes) required to pass this threshold varies depending on the power and price of current computing hardware, the nature of the data, and the type of processing required. For example, searching for the shortest road in a transport data set might be possible on a single computer, but computing an optimal route around a set of cities using the same data is a harder computational task which may require parallel processing and thus be considered to be a "big data" task.

Before the "big data" movement, classical database design strongly emphasized strict database design via ontologies, or formal ("neat") descriptions created in consultation with the client, specifying what is in the world that is to be represented by the data. Some "big data" proponents have argued that this is no longer the case and that the new era of data is about "scruffy" representations which can be reinterpreted by different users for different purposes. While this book covers both sides of this debate, it considers that there is still much merit in classical ontological ideas. It is possible that they may get "deconstructed" by the big data movement, but also that they will be "reconstructed" or rebuilt on new foundations created by that movement. In particular, this book covers the classical SQL database language and some recent attempts which rebuild it on top of "big data" tools. It considers that classical ontology is still relevant even in "messy" environments, as individual analysts must still use the same concepts in their individual interpretations of the data as a classical database administrator would use across the whole database.

Some critically important topics for real-world Data Science are rather less glamorous than the machine learning and visualization work usually associated with it in public. Data cleaning, database design, and software testing necessarily form a large part of related work. This book does not shy away from them but discusses them in detail consummate with their importance. Where possible, it tries to liven up these potentially dry topics by linked them to relevant ideas from Computer Science (Chomsky languages and data cleaning), Philosophy (ontology), and History (of the modern Calendar for date formatting). Some of these connections are a little tenuous but are intended to add interest and aid memory in these areas.

Readers following the mathematical details are assumed to be familiar with first-year undergraduate applied maths such as vectors, matrices, eigenvectors, numerical parameter optimization, calculus, differential equations, Gaussian distributions, Bayes' rule, covariance matrices. All these are covered in maths methods books such as the excellent,

- Riley, Hobson and Bence. *Mathematical Methods for Physics and Engineering (3rd edition): A Comprehensive Guide.* Cambridge University Press 2006.

If you are studying Transport Data Science to build your career, try searching at *www.cwjobs.co.uk* and other job sites for "data scientist" job descriptions and salaries to learn what types of roles are currently in demand. A fun exercise is to apply your new Transport Data Science skills to automatically process such jobs data; for example, you could try scraping salary, skills, and locations from the Web pages and drawing maps showing where certain skills are in highest demand. Then, link them to similar data from estate agent and transport data to visualize relationships between the salaries, house prices, and commute times to find your optimal job and location.

Please check this book's web site for updates or if you are having difficulty with any of the software exercises or installations.

Readers who enjoy self-studying from this book are encouraged to join the live Leeds course afterward, where they will meet many like-minded colleagues, work on the real project together, and build a strong personal network for the rest of their Transport careers.

Many thanks go to Neill Bennett and Dean Findlay at Derbyshire County Council for their help with content and application ideas, as well as for their forward-looking policy of open data which allows many of the examples in this book. From ITS Leeds: Richard Connors for help with content, structure, and style; Oscar Giles and Fanta Camara for checking and testing; my students Lawrence Duncan, Aseem Awad, Vasiliki Agathangelou, and Ryuhei Kondo and teaching assistant Panagiotis Spyridakos for their feedback; Greg Marsden, Richard Batley, and Natasha Merat for proving a stimulating environment for teaching and writing. Robin Lovelace and Ian Philips at Leeds and John Quinn at the United Nations Global Pulse team for GIS ideas. Subhajit Basu at Leeds' School of Law for ethical and legal ideas. Darren Hoyland at Canonical for help with OpenStack; Mark Taylor at Amazon for help with cloud services; Steve Roberts at Oxford for teaching me machine learning; Stephen $\mathcal{NP}$ Smith at Algometrics for teaching me big data before it was a "thing"; Andrew Veglio at Vantage Investment Advisory for teaching me "agricultural" SQL style; Thomas Hain at Sheffield for SGE; Peter Billington at Telematics Technology for the M25 project; Richard Holton for showing me the road from Athens to Thebes, Jim Stone at Sheffield for help with lucid pedagogy; Adam Binch at Ibex Automation for deep learning tests; To Jenny, if you are reading this I hope it is after I've read yours.

Leeds, UK   Charles Fox

Endorsements

"Transport modeling practice was developed in a data poor world, and many of our current techniques and skills are building on that sparsity. In a new data rich world, the required tools are different and the ethical questions around data and privacy are definitely different. I am not sure whether current professionals have these skills; and I am certainly not convinced that our current transport modeling tools will survive in a data rich environment. This is an exciting time to be a data scientist in the transport field. We are trying to get to grips with the opportunities that big data sources offer; but at the same time such data skills need to be fused with an understanding of transport, and of transport modeling. Those with these combined skills can be instrumental at providing better, faster, cheaper data for transport decision- making; and ultimately contribute to innovative, efficient, data driven modeling techniques of the future. It is not surprising that this course, this book, has been authored by the Institute for Transport Studies. To do this well, you need a blend of academic rigor and practical pragmatism. There are few educational or research establishments better equipped to do that than ITS Leeds."

—Tom van Vuren, *Divisional Director, Mott MacDonald*

"WSP is proud to be a thought leader in the world of transport modelling, planning and economics, and has a wide range of opportunities for people with skills in these areas. The evidence base and forecasts we deliver to effectively implement strategies and schemes are ever more data and technology focused a trend we have helped shape since the 1970's, but with particular disruption and opportunity in recent years. As a result of these trends, and to suitably skill the next generation of transport modellers, we asked the world-leading Institute for Transport Studies, to boost skills in these areas, and they have responded with a new MSc programme which you too can now study via this book."

—Leighton Cardwell, *Technical Director, WSP*

"From processing and analysing large datasets, to automation of modelling tasks (sometimes requiring different software packages to "talk" to each other, to data visualization, SYSTRA employs a range of techniques and tools to provide our clients with deeper insights and effective solutions. This book does an excellent job in giving you the skills to manage, interrogate and analyse databases, and develop powerful presentations. Another important publication from ITS Leeds."

—Fitsum Teklu, *Associate Director (Modelling & Appraisal) SYSTRA Ltd.*

"Urban planning has relied for decades on statistical and computational practices that have little to do with mainstream data science. Information is still often used as evidence on the impact of new infrastructure even when it hardly contains any valid evidence. This book is an extremely welcome effort to provide young professionals with the skills needed to analyse how cities and transport networks actually work. The book is also highly relevant to anyone who will later want to build digital solutions to optimise urban travel based on emerging data sources."

—Yaron Hollander, *author of "Transport Modelling for a Complete Beginner"*

Dr. Charles Fox is a University Academic Fellow in Vehicle and Road Automation at the Institute for Transport Studies, University of Leeds. He researches autonomous off-road, on-road, and road-side perception, control and data analytics systems, using primarily Bayesian methods. Recent projects include IBEX2 off-road autonomous agricultural vehicles, featured in The Times and on the Discovery Channel; INTERACT pedestrian detection analytics for autonomous vehicles, with BMW; UDRIVE data mining of manual car driving big data to identify causes of dangerous driving, with Volvo; and Automated Number Plate Recognition analytics for Mouchel and the Highways Agency. Dr. Fox holds a first class MA degree in Computer Science from the University of Cambridge, MSc with Distinction in Informatics from the University of Edinburgh, and a DPhil in Pattern Analysis and Machine Learning from the Robotics Research Group at the University of Oxford. He has worked as a researcher in robotics and data-driven speech recognition at the University of Sheffield for users including the BBC, NHS and GCHQ, and as a high frequency data-driven trader for London hedge fund Algometrics Ltd. He has published 50 conference and journal papers cited 800 times and has h-index 15. He is a director of Ibex Automation Ltd which advises hedge fund and venture capital clients and is available for consultancy and R&D work in Data Science and Robotics.

Contents

1	**"Data Science" and "Big Data"**		**1**
1.1	Transport Data Science Examples		1
	1.1.1	Origin-Destination Analysis on the London Orbital Motorway	1
	1.1.2	Airline Pricing and Arbitrage	4
	1.1.3	Pothole Monitoring	4
	1.1.4	Foursquare	5
	1.1.5	Self-driving Cars	5
	1.1.6	Taxi Services	5
1.2	The Claim		5
1.3	Definitions		7
1.4	Relationship with Other Fields		7
1.5	Ethics		10
1.6	Cynical Views		10
1.7	Exercise: *itsleeds* Virtual Desktop Setup		11
1.8	Further Reading		13
1.9	Appendix: Native Installation		14
2	**Python for Data Science Primer**		**15**
2.1	Programming Skills Check		15
2.2	Programming Languages		17
2.3	Programming Environment		17
2.4	Core Language		18
	2.4.1	Lists	19
	2.4.2	Dictionaries	20
	2.4.3	Control Structures	20
	2.4.4	Files	20
	2.4.5	Functions	21
2.5	Libraries		21
	2.5.1	Modules	21
	2.5.2	Mathematics	22
	2.5.3	Plotting	23
	2.5.4	Data Frames	23
	2.5.5	Debugging	24
2.6	Further Reading		24

3	**Database Design**		27
	3.1 Before the Relational Model		27
	3.2 Picturing the World		28
		3.2.1 Ontology	29
		3.2.2 Philosophical Ontology	29
		3.2.3 Data Ontology	34
		3.2.4 Structured Query Language (SQL)	37
	3.3 Exercises		37
		3.3.1 Setting up PostgreSQL	37
		3.3.2 SQL Creation Language	37
		3.3.3 SQL Query Language	38
		3.3.4 SQL Python Binding	40
		3.3.5 Importing Vehicle Bluetooth Data	40
	3.4 Further Reading		41
4	**Data Preparation**		43
	4.1 Obtaining Data		43
	4.2 Basic Text Processing		44
	4.3 Formal Grammar: The Chomsky Hierarchy		44
		4.3.1 Regular Languages (Type 3)	45
		4.3.2 Context-Free Languages (Type 2)	46
		4.3.3 Beyond CFGs (Types 1 and 0)	47
	4.4 Special Types		48
		4.4.1 Strings and Numbers	48
		4.4.2 Dates and Times	48
		4.4.3 National Marine Electronics Association (NMEA) Format	50
	4.5 Common Formats		50
	4.6 Cleaning		51
	4.7 B + Tree Implementation		52
	4.8 Exercises		53
		4.8.1 Reading the Database with Pandas	53
		4.8.2 *printf* Notation	53
		4.8.3 DateTimes	54
		4.8.4 Time Alignment and Differencing	54
		4.8.5 Parsing	55
		4.8.6 Vehicle Bluetooth Munging	55
	4.9 Further Reading		55
5	**Spatial Data**		57
	5.1 Geodesy		57
	5.2 Global Navigation Satellite System (GNSS)		59
	5.3 Geographic Information Systems (GIS)		61
		5.3.1 Role of GIS System	61
		5.3.2 Spatial Ontology	62
		5.3.3 Spatial Data Structures	63

	5.4	Implementations	65
		5.4.1	Spatial Files	65
		5.4.2	Spatial Data Sources	65
		5.4.3	Spatial Databases	66
		5.4.4	Spatial Data Frames	66
	5.5	Exercises	66
		5.5.1	GPS Projections	66
		5.5.2	PostGIS	66
		5.5.3	GeoPandas	67
		5.5.4	QGIS Road Maps	68
		5.5.5	Plotting Open Street Map (OSM) Roads	69
		5.5.6	Obtaining OSM Data	70
		5.5.7	Bluetooth Traffic Sensor Sites	70
	5.6	Further Reading	71
6	**Bayesian Inference**	75
	6.1	Bayesian Inference Versus "Statistics"	75
	6.2	Motorway Journey Times	76
	6.3	Bayesian Inference	78
		6.3.1	Bayes' Theorem	78
		6.3.2	Legal Inference: A Pedestrian Hit-and-Run Incident	78
		6.3.3	Priors and Posteriors	79
		6.3.4	Road User Tracking	80
	6.4	Bayesian Networks	80
		6.4.1	Bayesian Network for Traffic Lights	81
		6.4.2	Bayesian Network for Traffic Accidents	82
		6.4.3	Reporting	83
		6.4.4	Car Insurance	84
	6.5	Priors and Prejudice	85
	6.6	Causality	86
	6.7	Model Comparison and Combination	89
	6.8	Exercises	90
		6.8.1	Inferring Traffic Lights with PyMC3	90
		6.8.2	Inferring Accident Road State Change with PyMC3	90
		6.8.3	Switching Poisson Journey Times	91
	6.9	Further Reading	91
7	**Machine Learning**	93
	7.1	Generative *Versus* Discriminative Vehicle Emissions	93
	7.2	Simple Classifiers	95
		7.2.1	Linear Discriminant Analysis (LDA)	95
		7.2.2	Nearest Neighbor	95
		7.2.3	Template Matching	96
		7.2.4	Naïve Bayes Classification	96
		7.2.5	Decision Trees	97

	7.3	Neural Networks and "Deep Learning"		99
		7.3.1	Parallel Computing Back-Propagation	102
	7.4	Limitations and Extensions		103
	7.5	Exercises		105
	7.6	Further Reading		106
8	**Spatial Analysis**			107
	8.1	Spatial Statistics		108
	8.2	Bayesian Spatial Models		109
		8.2.1	Markov Random Fields (MRF)	109
		8.2.2	Gaussian Processes (Kriging)	113
	8.3	Vehicle Routing		114
		8.3.1	Link-breaking	115
	8.4	Spatial Features		115
	8.5	Exploratory Analysis		116
	8.6	Scaling Issues		118
	8.7	Exercises		119
		8.7.1	Gaussian Processes in GPy	119
		8.7.2	Gaussian Process Traffic Densities	122
		8.7.3	Vehicle Routing with PostGIS	122
		8.7.4	Finding Roadside Sensor Sites	122
	8.8	Further Reading		123
9	**Data Visualisation**			125
	9.1	Visual Perception		125
		9.1.1	Colours	125
		9.1.2	Visual Attention	126
	9.2	Geographic Visualization (Maps)		131
		9.2.1	Traffic Flow Maps	133
		9.2.2	Slippy Maps	136
		9.2.3	Info-Graphics	137
	9.3	Exercises		141
		9.3.1	Web Page Maps with Leaflet	141
		9.3.2	Bluetooth Origin-Destination Flows	143
		9.3.3	Large Project Suggestions	144
	9.4	Further Reading		145
10	**Big Data**			147
	10.1	Medium-Sized Data Speedups		147
	10.2	Enterprise Data Scaling		149
	10.3	CAP Theorem		151
	10.4	Big Data Scaling		152
		10.4.1	Data "Lakes"	152
		10.4.2	Grid Computing	153
		10.4.3	Map-Reduce and Cloud Computing	154
		10.4.4	Hadoop Ecosystem	155
		10.4.5	Non-relational Databases ("NoSQL")	156
		10.4.6	Distributed Relational Databases ("NewSQL")	157

	10.5	Exercises	158
		10.5.1 Prolog AI Car Insurance Queries	158
		10.5.2 Map-Reduce on Vehicle Bluetooth Data	158
		10.5.3 Setting up Hadoop and Spark	160
		10.5.4 Finding Vehicle Matches in Hadoop	160
		10.5.5 Traffic Flow Prediction with Spark	161
		10.5.6 Large Project Suggestions	164
	10.6	Further Reading	164
11	**Professional Issues**		**165**
	11.1	Morals, Ethics, and Law	166
	11.2	Ethical Issues	167
		11.2.1 Privacy	167
		11.2.2 De-anonymization ("Doxing")	168
		11.2.3 Predictive Analytics	170
		11.2.4 Social and Selfish Equilibria	170
		11.2.5 Monetization	171
		11.2.6 Ontological Bias	173
		11.2.7 *p*-hacking	174
		11.2.8 Code Quality	176
		11.2.9 Agency Conflicts	177
		11.2.10 Server Jurisdiction	177
		11.2.11 Security Services	177
	11.3	UK Legal Framework	178
		11.3.1 Data Protection Act 1988	178
		11.3.2 General Data Protection Regulation (GPDR)	179
	11.4	Role of the Data Scientist	180
	11.5	Exercises	181
	11.6	Further Reading	181
Index			**183**

"Data Science" and "Big Data"

The quantity, diversity and availability of transport data is increasing rapidly, requiring new skills in the management and interrogation of data and databases. Recent years have seen a new wave of "Data Science", "big data" and "smart cities" sweeping though the Transport sector. Transportation professionals and researchers now need to be able to use data and databases in order to establish quantitative, empirical facts, and to validate and challenge their mathematical models, whose axioms have traditionally often been assumed rather than rigorously tested against data. In 2012, the Harvard Business Review described Data Science as "the sexiest job of the 21st century", and in 2011 consultancy McKinsey predicted demand for 1.5 million new jobs in Data Science. While the term with its current meaning has been in use since 1996, it only began to appear as a common Silicon Valley job title from around 2008, and is now a buzzword. The term "big data" is similarly omnipresent in the world's media, used by most journalists, though not by most academic researchers, as a synonym for "Data Science". What are these apparently new disciplines which have ascended so rapidly? And how much is hype which simply re-packages much older work in related fields such as Statistics and Computer Science?

Before attempting to define any of these terms this introductory chapter will consider some examples of transport applications which have recently been made possible by them. It then considers some of the claims made by Data Science proponents and opponents and offers some definitions. It then shows how to install a complete transport data science software stack on the reader's computer.

1.1 Transport Data Science Examples

1.1.1 Origin-Destination Analysis on the London Orbital Motorway

Our 2010 paper (Fox et al. 2010) re-purposed several months' of existing Automatic Number Plate Recognition (ANPR) and induction loop data from the M25 motorway around London, to estimate origins and destinations of the actual routes taken by drivers. ANPR detects the numbers and letters on each vehicle's number plate, while induction loops are pressure sensors built into the road which count the number (flow) of vehicles driving over them. The system provides a graphical tool, shown in Fig. 1.1, which allows transport planners at the Highways Agency[1] to click any origin and destination

---

[1] Now Highways England.

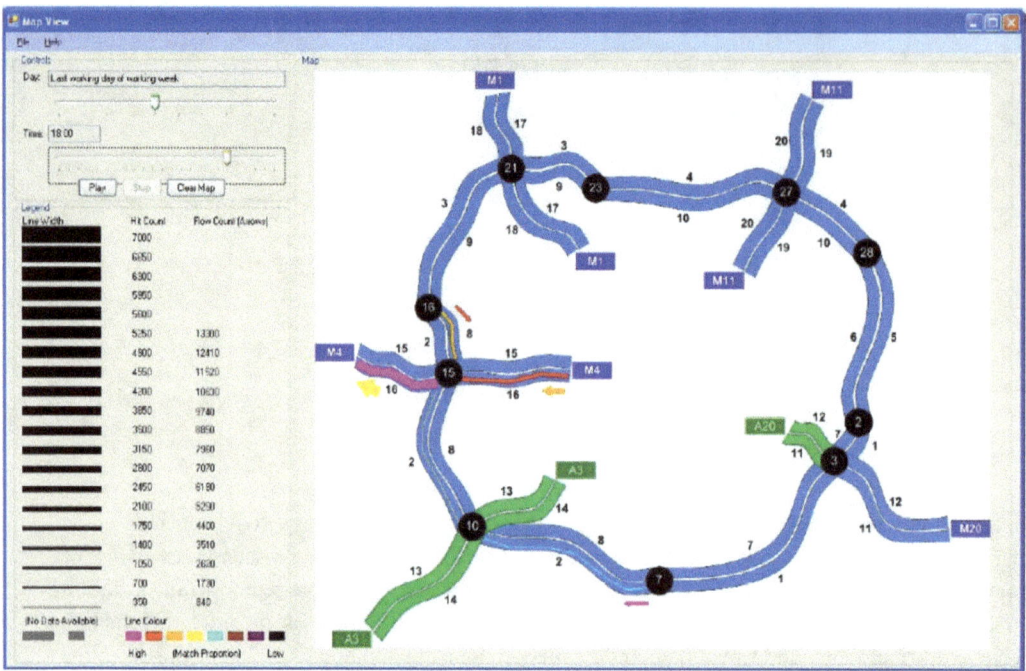

Fig. 1.1 Graphical User Interface (GUI) to a data-driven origin-destination system. The user selects an origin and destination by clicking on locations around the London M25 motorway, and receives an estimate of flow along this origin-destination as a function of time of day

on the network, to see an estimate of the number of journeys between that origin and destination as a function of time of day, as in Fig. 1.2. This enables them to make planning decisions such as where best to install smart management systems, add new capacity, or encourage public transport alternatives. In particular it was found that many journeys on the M25 are very short, consisting of only arterials and two motorway junctions, which suggested that new public or private capacity could be provided to connect their origin and destination towns directly to take this traffic off the motorway, freeing it up for longer journeys.

The user-facing tool conceals various machine learning and statistical inferences used to obtain the estimates. This ANPR data is "hashed" at source, i.e. the raw characters of the number plates are replaced by shorter strings of numbers inside the cameras, which is intended to make it impossible to identify individual drivers, and to enable only bulk analysis from them. The cameras on the M25 did not cover all of its lanes, being typically installed (for previous purposes) in pairs which covered only random sets of two out of the three motorways lanes at most locations. Together, these two problems mean it is non-trivial to reconstruct particular journeys. Vehicles could not initially be identified by unique number plates or linked to their owners, and many vehicles would seem to appear and disappear as they drove in lanes not covered by cameras. However by fusing data the system is able to a large extent to compensate for all this. By fusing a month of data from every ANPR detection on the M25 with corresponding induction loop flows, which measure the total number of vehicles across all lanes at locations, we could estimate the ratios of vehicles being lost from camera tracking, and adjust our bulk estimates of each origin-destination route's usage accordingly (Fig. 1.3). To calibrate parts of this process, it was necessary to go beyond pure Data Science and add a small, carefully hand-collected, data set as in "normal science". This was done by installing new experimental cameras at a few locations, with known parameters. By detecting ground-truth ANPR codes from these cameras, the

1.1 Transport Data Science Examples

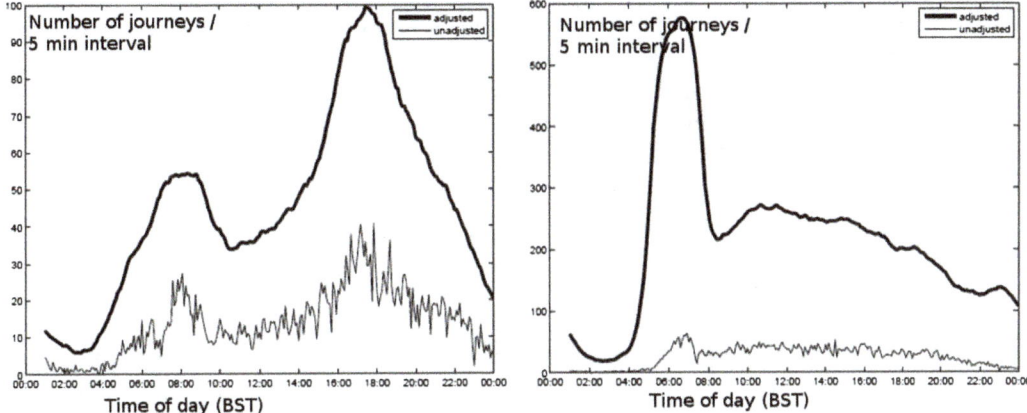

Fig. 1.2 Outputs from the M25 origin-destination analysis tool, for two routes. The graphs show the raw detected number of matching number plates at the original destination as a function of time of day; and adjusted versions which give the best known estimate of the true number of such journeys after various machine learning and Bayesian statistical operations to adjust for missing and noisy data

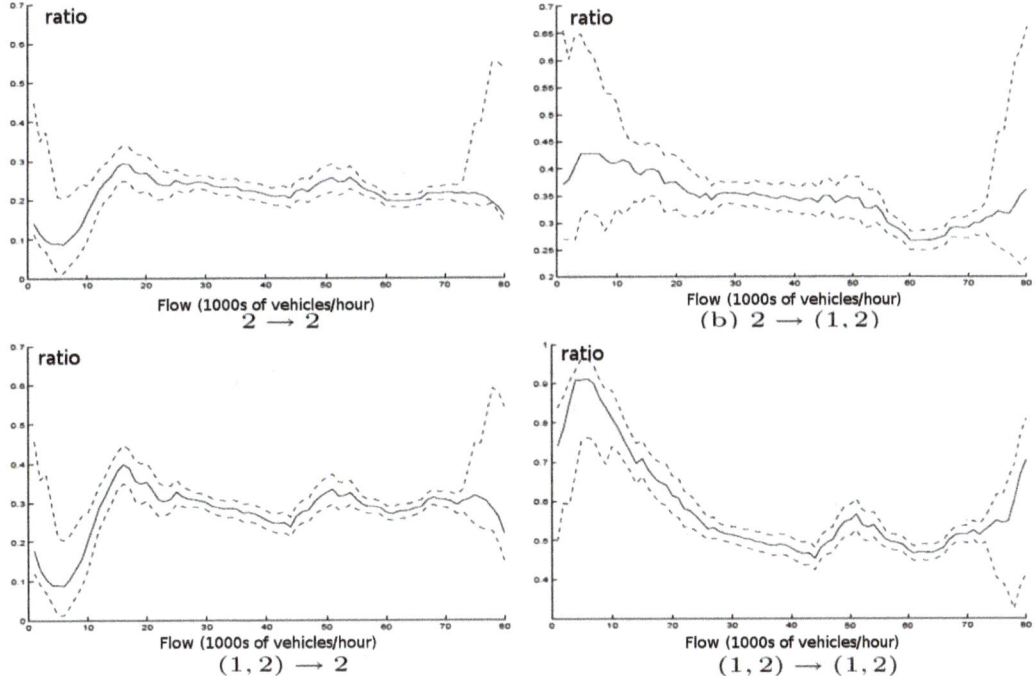

Fig. 1.3 Empirical detection ratios for partial camera coverage, used in the adjustments, as a function of total flow as measured by induction loops, and as a function of lane coverage type. Notation "*(1,2)->2*" means the setting where the origin motorway has three lanes with camera in lanes 1 and 2, and the destination is a three-lane motorway with camera only in lane 2

detected vehicles could then easily be tracked around the rest of the network from the hashed cameras, and estimates of detection losses due to missing cameras computed for various environments. As a side-effect, this also meant that we could track a large portion of journeys of identifiable individuals around the M25, from data that was supposed to be anonymous! This serves to illustrate an important point: even supposedly anonymized data can be "cracked" to regain the driver identities, in this case through a mixture of fusion with other sources and collection of additional experimental data.

1.1.2 Airline Pricing and Arbitrage

When is the best time to book your flight on-line? Dynamic pricing has long been used by airlines to adapt to supply and demand. In particular, the last few seats on a plane may be cheap because the plane is scheduled to fly anyway and needs to fill them; but booking one more trip after a plane is filled would require a larger plane to be rescheduled in its place, or another journey planned. According to some reports, the best time to book is 137 days before the flight. Some people and companies make a living by arbitraging the time values of flight tickets, buying and selling them to help to balance out demand. Data scientists will draw graphs of demand over time, and try to isolate factors within it such as business/pleasure travel, demographics, and wealth of the customers. Because customers are already accustomed to dynamic pricing, it may be easy to apply differential pricing to customers based on their wealth. Data companies such as social networks and search engines have collected decades of personal information on individual users which can be used to build detailed psychological and financial profiles of them. This data may be sold to data aggregation companies, who in turn sell it on to the highest bidders such as airlines. An airline could examine your profile and decide how much you are likely to be prepared to pay for your flight according to your psychological and financial profile. If they learn from your social media profile that you are a highly paid transport professional and are booking a flight to present your work at the location of a known Transport Studies conference, then they can infer that this is an important and time-critical trip for you which can be priced highly. This is how data companies make much of their money while offering apparently "free" on-line services. Historically, (and mirroring many other areas of finance) we have seen these ideas appear first via dedicated arbitrage companies, set up to make money by automatically buying and selling tickets. Later, the suppliers themselves notice this happening, and implement their own prediction systems to try to take the profit back from the arbitrage companies. Suppliers may have options available to them that arbitrageurs do not, namely the ability to bring new real capacity on-line, and to use to this alter prices.

1.1.3 Pothole Monitoring

Locating damaged road surfaces has historically been a time consuming processes, requiring local authorities to send out survey vehicles and manual spotters who log the locations of potholes and other damage. Other authorities rely on citizens reporting them in their local areas, but this can lead to a distortion where areas with the most active citizens get their roads fixed more than those with less active citizens. Using cheap sensors and communications, it is becoming possible to conduct full surveys of road networks, in some cases almost for free. In Boston, researchers used a mobile phone app given to citizens to log acceleration data from the inertial measurement units (IMU) of their phones, and send it to the local authority. When driving over a pothole, the vertical accelerations can be detected and classified as a pothole in need of repair. In other cases, local authorities own vehicles, which drive around their areas for other purposes, have been retro-fitted with lidar (Light Imaging Detection And Ranging) laser scanners which build 3D maps of the surface and make similar classifications and logs.

1.1.4 Foursquare

Foursquare is a phone app which allows users to "check in" to urban locations such as restaurants and bars, review and recommend them, and locate their friends. This type of urban spatial data could be valuable for transport planning, as it allows planners to understand the end destinations of some journeys, and watch how their patterns change over time. As a new area becomes fashionable for restaurants, planners might respond by providing more public transport, or altering road signals. At the micro level, predictions could be made about individual road users' behavior, for example someone who often drives to the same bar on Friday evening could be predicted to be going there again, and this information fed into road signal optimizations to help them to get there.

1.1.5 Self-driving Cars

Current autonomous vehicles are very dependent on data, specifically on high-resolution 2D and 3D maps of the large areas in which they operate. The JUNIOR Urban Challenge vehicle used a map containing individual locations of curbs and white lines on the road which it used to navigate. Historically, self-driving car research has driven much spatial database technology, for example the original 1980s experiments on automation at Carnegie Mellon University gave rise to predecessors of the "simple features" spatial databases used in this book.

1.1.6 Taxi Services

Both manual and eventually autonomous private hire vehicles will collect data about passengers and drivers, which could be linked with other personal information. As with airlines, taxi companies could buy up financial data on individuals from social networks and mix it with surge pricing to mask the use of differential pricing to charge higher fees to rich and needy people. Taxi companies could also sell their own data on their customers' and drivers' reputations to data aggregators, for example insurance and loan companies might be interested in both passenger and driver reputations. As with flight arbitrage, we are starting to see trickle-down effects where first new digital arbitrageurs – who might not operate any vehicles at all – and then established taxi companies and even public sector local authorities deploy demand predictions. They may be especially useful for authorities such as Derbyshire County Council which currently must operate scheduled rural bus services as a social service. Such services provide important psychological assurance that all citizens *could* travel around the area if and when required, even when many buses often have no passengers. Replacing them with a socially operated taxi service, optimized by data, could save on money and emissions while retaining and even improving (providing door to door rather than scheduled stops) the social service.

1.2 The Claim

What do these transport applications have in common that makes them "Data Science" or "big data"? According to Mayer-Schonberger and Cukier's influential characterization, "big data" is about three fundamental changes to the scientific method:

Bigness. In "normal science", collecting data is expensive as requires design and performance of experiments. Hence it uses Statistics to estimate the size of data required to achieve significant results, and uses sampling and estimation methods to obtain them. In contrast, Data Science now has "the ability to analyze vast amounts of data about a topic rather than being forced to settle for smaller sets".

The effects of this is that statistical sampling and estimation become unnecessary, and that models with very large numbers of parameters can be fitted to data without testing for or worrying about over-fitting. It is sometimes possible to work with all of the data rather than any sample from it. This has arisen for several interconnected technological reasons over the past decade. Sensor technology has fallen in price, allowing more data to be collected. For example, rather than employing people to count cars in traffic surveys, we can now cheaply deploy large networks of traffic cameras. It is possible to buy web-cams for a few pounds now rather than hundreds or thousands of pounds. Often the most expensive part of a sensor nowadays is the metal enclosure to keep it waterproof. Computation power has fallen in price, allowing more data to be processed. This includes both statistical analysis of large data sets, as well as enabling sensor data to be processed into database entries in the first place. For example, ANPR algorithms have been known for decades but can now run cheaply on embedded processors inside traffic cameras in a network. The Raspberry Pi and Arduino are examples of cheap computing boards which can be built into sensors' boxes for less than 100 GBP. Data storage has fallen in price. Most computer owners now have terabytes of storage space available, which was previously only available to large companies. Data connectivity has increased. The internet has moved from being a static collection of web pages to an interactive space where organizations routinely publish and update raw data sets which anyone can access. Internet speeds have increased and the cost of bandwidth has fallen to enable anyone to download large data sets easily. Internet-based hosting allows home users to store and process data remotely ("in the cloud") for a few pounds.

Messiness. "A willingness to embrace data's real-world messiness rather than privilege exactitude". Statistics, Database design, and Artificial Intelligence have historically focused on working with well-defined and structured "clean" data. "Ontology" is the study of defining what exists in a model, and its role is reduced in Data Science. Often, Data Science algorithms will run directly on raw, unprocessed data files directly downloaded from some public provider, and bypass the classical architecture of database design and insertion altogether. Data Science works with whatever data is available, to make it as large as possible. Usually this will be data collected for some other purpose. Because data is so cheap nowadays, we can simply make use of very large collections of noisy data and average out over the noise, rather than have to carefully design and collect clean data. For example, rather than design samples of manual vehicle counting locations and strict manual protocols for classifying the vehicle types, we make use of a city's existing large network of ANPR sensor data. Every ANPR detection and vehicle type classification (which can be done with machine vision or vehicle registry lookup) is noisy but there is so much of it that it no longer matters. Data Science tends to emphasize quantity over quality, and the reuse of whatever second-hand data it can find. Often this means that the data scientist has to do more work in "munging" or pre-processing one or more data sets into a usable format.

Correlation. "A growing respect for correlation rather than a continuing quest for elusive causality". In "classical" science and statistics, careful distinctions are made between causation and correlation. Causation is a difficult philosophical concept, though has recently been given solid statistical foundations. Causality can typically only be inferred from a system when causality is put into the system. This occurs in controlled scientific experiments, where the experimenter forces some variable to take a value and records the result. In contrast, Data Science is a passive activity, which analyses only data which already exists and has not been caused by the data scientist. Hence, Data Science deals with correlations rather than causation. This means working with "black box" parametric models rather than with generative, theory-driven models. Data scientists argue that the predictions generated by black boxes are often better than those of theory-driven models, which is all that matters.

Other authors have used related terms, the "4 V's" to characterize "big data": "velocity, variety, volume and veracity" to capture similar concepts. Here, "velocity" emphasizes real-time updating and speed of arrival of new data, "variety" means its messiness and lack of formal structure, "volume" its size, and "veracity" (or rather, lack of) refers to the noise in the data and lack of guarantee of its consistency and truth.

1.3 Definitions

In this book, we separate out the concepts of "Data Science" from "big data" as follows:

Data Science here means the use and re-use of data collected passively rather than via the causal experiments used in regular "science". Causal inference has recently (from the early 2000s) become well understood within the framework of Bayesian networks. It is discussed in more detail in Chap. 6, but roughly we now know that, contrary to the claims of some earlier statisticians and educators, it is possible to infer causation and not just correlation using statistics, but (except for a few special cases) only if the data itself has first been caused in some way by the experimenter. For example, to learn whether mobile phone use causes car crashes, it is necessary to cause some drivers to use their phones and others not to use them, and observe the results, rather than just observing a set of drivers' phone and crash behaviors. Without this kind of experimental control, it would be possible for confounding factors such as personality to cause both phone use and crashes, and thus only possible to infer their correlation. Data Science is thus unambiguously defined and cleanly distinguished from regular "science" as science without causation. This concept corresponds to "correlation" above. While data scientists could collect their own data, they typically work with existing data gleaned from various sources such as previous experiments, the internet, company databases and government records. This practicality gives rise, incidentally, to the "messiness" above, and sets the character for much of practical Data Science's data preparation or "munging" activities.

Big data here means data that cannot be usefully processed on a single computer processor, and which is therefore processed using parallel computation. Parallel computation is a traditional and well-known area of Computer Science, with roots stretching back to the "parallel Turing Machine" in the 1930s. However, "big data" is not synonymous with "parallel computing" because parallel computing is more general, for example it may be used to perform detailed physical simulations or compute solutions to pure mathematics problems, neither of which involve the processing of any data. Our definition is relative to usefulness in a user's application, so depends on the application as well as the data's size. For example, summing the length of the roads in a network might not be a big data problem, but finding the shortest path between all the cities in it (the "Traveling Salesperson Problem") may be one, because it requires much more computation power. (Some authors would consider our definition to be too general and reserve "big data" for the processing of data the scale of the Silicon Valley giants – whole data center buildings of exabytes of data. However it is very hard to pin down a meaningful definition of how big this needs to be. The switch from serial to parallel processing in contrast provides a very clear and definable point. Parallel computing can mean making use of the four cores of your laptop together, using your GPU as a compute engine, using 100 GPUs in a cluster, or using 100,000 commodity PCs in a data center.)

1.4 Relationship with Other Fields

The terms "Data Science", "big data", "data analytics" are often used apparently interchangeably, sometimes also with "machine learning", "data mining" and "statistics". None of these terms have widely accepted definitions – there are no standards organizations determining who is a "data scientist", and generally having any of them on your CV will get you considered as being one in the current job market. Rather, they are all generally used as fuzzy concepts to label rough, overlapping, clusters of ideas and activity, though with different emphases. Some of the characteristics of these emphases are listed below.

- *Classical Statistics.* The notion of a "statistic" comes from Classical (non-Bayesian) Statistics. A "statistic" is a function of some data which may be used to estimate the value of an unobservable characteristic of the population from which the data was sampled. Classical Statistics invents definitions of statistics then proves their properties such as bias as convergence speed as estimators. Some statistics may be shown to be optimal, or better than others, in these senses. Other statistics might not estimate anything causal or theoretic, but be invented just to describe some interesting property of the data, or to describe measures of confidence in the results of estimators (such as p-values). Statisticians are consulted to design experiments to collect new data.
- *Bayesian Statistics.* Is a mis-named field, better called "Bayesian inference", because it does not generally deal with "statistics" at all in the classical sense. It does not invent functions of data which are used to estimate things. Instead, Bayesians begin with one or more generative parametric models which they assume have given rise to the observed data. Rather than estimate the values of the parameters of these models, they infer them probabilistically, using Bayes' rule to invert the generative models. Bayesians do not use p-values or confidence intervals. Instead, they ask what is the posterior belief about the value of the generating parameters given the data, which they ideally report as a full probability distribution over them. Doing Bayesian Statistics properly is generally computationally "hard" (in the sense of $\mathcal{NP}$-hard) which means that it cannot usually be done exactly in practice. Instead, Bayesians use a mixture of approximation algorithms and brute-force computation power to report approximate results.
- *Econometrics* studies particular types of statistics for use in economics, especially time-series modeling and prediction.
- *Operations Research* emphasizes actions and utility, and multi-party interactions, using statistics and data.
- *Database Administration (DBA)* has existed as a specialist career for decades, and can include the design of database structure, preparation and insertion of data into the database from various sources, daily maintenance of computer systems and provision of access to data by users, as well as asking questions of the data. In small companies this work might overlap with more general IT (information technology) work such as system administration or website design. Many business websites are essentially front-ends to databases of product and sales information, together with tools enabling users to place orders.
- *Database Systems.* Computer Science includes this study and design of algorithms and software to implement database tools themselves. This becomes very complex when databases run across many computers, serving many users concurrently, and distributing both computation and storage over nodes and networks as in true big data systems.
- *Data Ontology.* The initial design of databases can be a huge business and philosophical process, requiring detailed conceptual understanding of business process as well as general "ontology" issues to choose what entities exist is the world and how to model them. Philosophical ontologists spend centuries debating what exists. Data ontologists must be quicker and messier, but do address some of the same questions. This work involves business consulting skills to work with business process managers to understand and model their concepts.
- *"Data analysts"* or *"business analysts"* are focused on using the database to formulate and answer more detailed questions, which may involve some element of statistics in their design, as well as business consulting skills to understand upper management's questions and translate them into database and statistical problems. Typically an analyst will be called by a manager with a vague question like "where could we invest in building a new road to reduce congestion?", and the analyst will work with data and statistics to produce a report full of evidence about what decision to take.

1.4 Relationship with Other Fields

- *"Data analytics"* or just *"analytics"* describes the work performed by data analysts, though with an emphasis on the technical rather than business work. A "data analytics" group might have more autonomy to conduct its own exploratory research than a "data analyst" working for a business manager. For example, a transport data analytics team might have a budget to commission its own data collection and explore its own hypotheses about what is happening on a traffic network, and suggest new transport policies, rather than just answering questions set by non-technical transport managers.
- *"Data Mining"* is used to refer specifically to the use of non-generative, black-box models of data, and often to exploratory studies in particular. For example, we might not have any particular question to ask of traffic flow data in a city, but we have the data available and would like to get some idea of "what is going on" which might generate theories and hypotheses later on. Or we might want to predict the traffic flow at a location from the data without any need for a generative or causal model. Sometimes the derogatory term "data dredging" is using to emphasize the well-known problems of over-fitting and spurious correlations that can plague data mining.
- *"Data visualization"* is the art and science of presenting results of data exploration, or more theory-driven studies, in visual form. The media call it "data journalism". This can overlap with visual psychology and cognitive studies, and with graphic design and art.
- *"Machine learning"* originally referred to the academic study of black-box data-mining algorithms, for example proving theorems about their convergence properties. However it is now used more generally to mean the use of data mining, pattern recognition, and also of generative computational Bayesian models. Historically, its academic study included non-statistical methods of symbolic concept learning, though most work now is statistical.

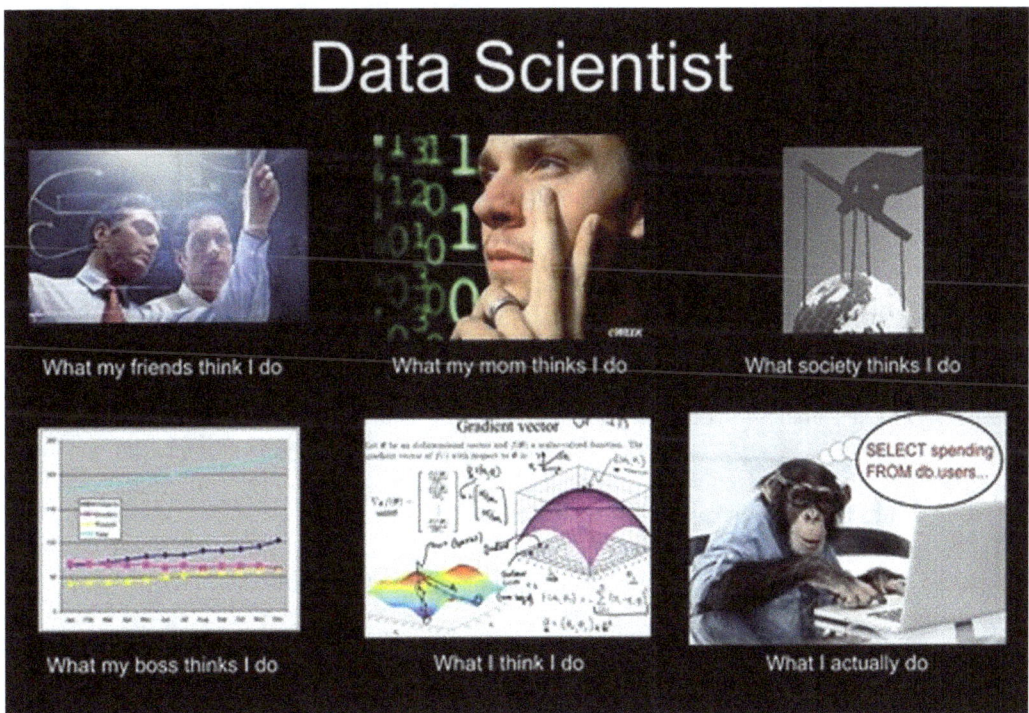

Fig. 1.4 Internet "meme" (*anon.*)

- A *data scientist* is probably someone who has interests in some or all of the above areas. As a culture, Data Science also tends to emphasize practical skills over theory, and the use of and contribution to open source software. There is an active community of open source users and developers based around standard tools, which a data scientist is usually expected to know. You will be introduced to many of these in this book. Data scientists often enjoy sharing data, code, results, and visualizations on the Internet, and believe that the world can become a better place through this process (Fig. 1.4).

1.5 Ethics

Most people feel that the use of aggregate data from collections such as the above examples is "OK" but begin to worry when the same data is used to make predictions about, and to act upon, individuals. For example, we might fuse data from ANPR and supermarket loyalty reward cards to determine that customers who drive to supermarkets buy larger packs of drinks than those who walk there, and help both types of customers to find the products they desire more efficiently by placing them near separate car park and high street entrances. However, the same data sets could be fused to locate all the pairs of individuals who hold reward cards with partners and regularly buy nappies and milk, but regularly drive to a hotel on Friday nights with non-partners having bought wine and high-end underwear. This data could then be sold, along with names and addresses of all concerned, to the highest bidder to profit from as they choose. Legal codes exist to regulate data use, though in practice many people just click "OK" rather than reading detailed legal policies of many data collectors. We will discuss ethics and law in more detail later in the last chapter.

1.6 Cynical Views

Data scientists are undoubtedly highly sought after and highly paid, both in transport and more generally. Is Data Science a new revolutionary approach to science and life, or just a fad or a re-branding of older ideas? Recent years have seen something of a data backlash, with criticism coming from several angles:

- "After a hard day's Data Science I like to do some Food Cooking and some Plant Gardening". "Real scientists" have pointed out that *all* science uses data and has done so for many centuries. CERN and similar experiments have produced "big data" for decades, and physicists use statistical and computational methods to analyse it. Meta-analyses collect results from multiple previous studies to fuse information from many data sets collected by other scientists. Transport scientists already commission large scale studies to collect and analyse traffic flows and origin-destination analyses, and commuter preferences. Every science thesis ever written has involved examining some data, and computing and reporting some statistics of it. Transport modelers, psychologists and economists have all been using statistical packages such as Excel, Matlab and S-Plus for many decades. So what justifies self-describing "data scientists" being anything other than "scientists"? Can any kid who has ever computed the mean of a spreadsheet column in Excel now call themselves a "data scientist" or "big data expert"?
- "Real scientists make their own data." Here the "real scientist" does make a distinction between Data Science and "real science". The claim is that Data Science is simply a subset of Science, being that which works only with old data rather than having the ability to also collect new data. "Real science" is about the careful design and execution of experiments, which produce (usually) small but significant results through careful design and sampling. Real science is harder than Data Science

because it requires these additional skills. Data is not messy because it has been collected with care and skill. Most importantly, real science can make claims about causation because it collects its own data. Data Science can never provide a "true" theory of reality, it only shows correlations. These may be useful in practice, such as predicting when to impose speed limits on motorway lanes, but will not provide the intellectual insight of a generative, causal model of the underlying psychology and economics of human driver behavior.

- "Real statisticians" have pointed out that use of big data is often unnecessary. Just because you have 100 million number plate detections collected does not mean that you will get a better understanding of origin-destination routes than the conventional approach of taking a smaller but carefully selected sample that is designed to achieve statistical significance. The additional data carries no additional useful information beyond this point.
- "Real statisticians" also point out that the use of big data can also be dangerous, and wrong, due to sampling effects. A core assumption of big data is that there is so much of it that sampling and estimation are no longer needed. If *all* possible data is available (sometimes written as the slogan "$N =$ all"), then this is correct. But in practice most "big" data sets are still incomplete for most uses, despite their "bigness". For example, predictions for the 2016 USA election and UK Brexit votes made heavy use of social media data, assuming that is was so big that it represented the whole countries' populations. But social media users are disproportionately young and rich compared to the whole population, resulting in large biases in the samples. Data may be big but still be only a sample, in which case the old statistical methods are still needed.
- "Crossing the creepy line". Businesses have their own reasons for promoting new technologies to customers and investors, which may differ from motivations of academic researchers and governments. Some have argued that the large sums of money invested in the area by large companies has acted to distort the research field, favoring research in more profitable directions (specifically, prediction of advertisement clickthroughs and shopping basket correlations), and encouraging the use and abuse of personal data. The flip-side of Data Science's belief in sharing and progress is a general disregard for individual's privacy, as they seek to extract as much knowledge as possible from whatever data is available. Similarly, where governments have collected data on citizens, such as transport commuting patterns, there may be scope to go beyond aggregate statistics and intrude into individuals' behaviors.

1.7 Exercise: *itsleeds* Virtual Desktop Setup

All the software used in this book is open source and will always be available free of charge,[2] both for your studies and for commercial or any other use afterwards.

itsleeds is a standardized virtual desktop which accompanies this book and is also used to teach ITS Leeds students. It automatically installs all of the tools, commands, and libraries referred to throughout the book, on almost any modern computer so that all readers can work with the exact same computer setup. To use it, you must first install either the program Docker from *docker.com*, or for some lower capability computers (at the time of writing, these include computers running Microsoft Windows 7, 8 and 10 Home; and Apple Macs from before around 2010) the program Docker Toolbox from *docs.docker.com/toolbox/overview*. (Some systems may require the VirtualBox install option to be

---

[2] "Open source" means that the source code of the software is available for anyone to inspect and modify, as well as to run free of charge. "Free software" is a stronger and more political term which refers not to price but to "freedom", and which ensures in addition that modifications of code must to contributed back to the community rather than sold. See *www.fsf.org* and *www.catb.org* for details.

selected during setup, and any previously installed virtual machine systems to be removed prior to installation.) If you have any problems with installation, please check this book's website for updates.

A copy of *itsleeds* has been placed on Docker's central server, where its full name is *itsleeds/itsleeds*, and Docker will automatically download it given this name. It is a gigabyte sized file which may take some time to download. The required Docker command is,

```
docker run -p 6901:6901 itsleeds/itsleeds
```

For most Microsoft Windows setups: type this command in the Docker Quickstart program. For most Apple Mac and Linux setups: use a terminal (*Applications*→ *Utilities*→ *Terminal* on OSX). Linux and Mac users may need to prefix this and all other Docker commands with *sudo* such as,

```
sudo docker run -p 6901:6901 itsleeds/itsleeds
```

On some computers you may need to run *docker* tools as "Administrator", "root", or similar privileged modes.

Then open the *itsleeds* virtual desktop inside your web browser, by navigating to its address. For Mac and Linux setups this is usually,

```
http://127.0.0.1:6901/?password=vncpassword
```

For most Microsoft Windows setups, Docker Quickstart will print a number (an IP address) when it first starts up with a format similar to 127.0.0.1, *before* any Docker commands are given. Make a note of this number and use it in the address instead of 127.0.0.1, for example,

```
http://123.456.789.012:6901/?password=vncpassword
```

You should see a desktop as in Fig. 1.5 inside your browser. *itsleeds* behaves as if it was a separate physical computer on your network. It can access the internet, e.g. using its web browser. A simple way to move files between it and your real computer is just to email them to yourself!

In the rest of this book we assume that you are working inside the *itsleeds* virtual desktop. In the text, commands which are preceded by the prompt "$" (including running *psql*) are for use in the terminal command line.[3] Use the *Applications* → *Terminal Emulator* to obtain a terminal to type in. For example, open a terminal then type,

```
$ ls
```

to *l*ist the files in your home directory. Do not type the "$" symbol itself, and press return at the end of the command to execute it. Other commands in this book without the "$" are to run inside Python or *psql* according to context (and in this section only, Docker commands on your own real computer). Commands which are too long to fit on one line of text in the book are broken up using an backslash ("\"). Neither the backslash nor the newline ("return") should be typed, rather the whole command should be entered as a single unbroken line.

In this book we will use the graphical integrated development environment Spyder for Python work. It is included in the Docker image. To start it, go to *Applications* → *Development* → *Spyder3*.

The screen-shot in Fig. 1.5 shows the Docker image running inside a web browser. The first window shows a *psql* startup, and the second shows Spyder.

The *itsleeds* virtual desktop is represented on your local computer by a single file. If you make changes to the state of the virtual desktop, for example, by doing any work and writing to its *virtual* file system, then you will need to "commit" (save) the new state of the *itsleeds* file onto your real computer.

---

[3] The operating system used inside *itsleeds* is Ubuntu and the command line program is *bash*.

1.7 Exercise: *itsleeds* Virtual Desktop Setup

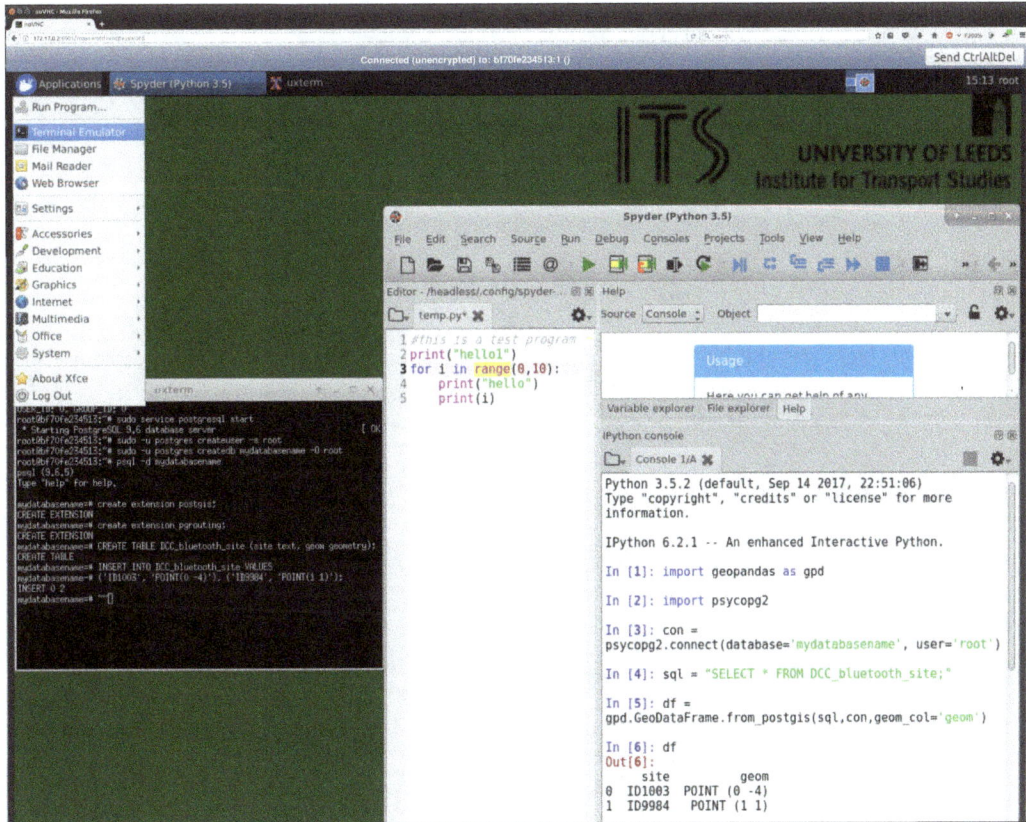

Fig. 1.5 The Docker virtual desktop *itsleeds*, running inside a browser, performing basic spatial data commands from later chapters in *psql* and *Sypder*

Check Docker's instructions for your own type of computer for how to commit and reload. Typically this is done by first obtaining an identifier code for the virtual desktop with a command such as,

```
docker ps -l
```

Then using this identifier code (in place of *<container_id>*) to tell Docker which virtual desktop (as Docker can run several at once) to commit to a new file, such as *itsleeds-myedit*,

```
docker commit <container_id> itsleeds-myedit
```

Then reload and run your modified virtual desktop *itsleeds-myedit* instead of the original *itsleeds* with,

```
docker run itsleeds-myedit
```

1.8 Further Reading

Some easy and fun popular science books which give some insight into the culture of Data Science,

- Mayer-Schonberger V, Cukier K (2013) Big data: a revolution that will transform how we live, work and think. John Murray Publications

- Townsend, AM (2014) Smart cities – big data, civic hackers, and the quest for a new Utopia. Norton
- Lohr S (2015) Dataism. One World Publications

Full details of the M25 motorway origin-destination model,

- Fox C, Billington P, Paulo D, Cooper C (2010) Origin-destination analysis on the London orbital automated number plate recognition network. In: European Transport Conference (Available on-line)

A UK government report on Transport Data Science,

- The transport data revolution. UK Transport Catapult report. Available from: https://ts.catapult.org.uk/wp-content/uploads/2016/04/The-Transport-Data-Revolution.pdf

1.9 Appendix: Native Installation

We will assume you are running on *itsleeds* throughout this text. However if you wish to install the tools natively on your own computer, at your own risk, some tips are provided in the *NativeInstallation.txt* file distributed with the Docker image. These are working at the time of writing, though versions may change in future. Configuration of some of the tools may be difficult for beginners, who are strongly recommended to use *itsleeds* rather than attempt their own installations. If you are a beginner and *really* want or need local installation then you should contact your IT support for assistance with the process.

Python for Data Science Primer 2

In this book we will use SQL and Python as the main programming languages in computational examples. This chapter introduces new programmers to Python. It may be skipped by readers who are interested only in concepts and management rather than the details of programming. If you already know Python and are able to complete the following skills check then you can also skip this chapter. If you are new to Python then try to complete the skills check with the help of the chapter and references before moving on. It is not possible to learn all of Python from a single chapter, but we provide a guide to the main parts relevant to Transport Data Science, which can be consulted in more detail in the references provided. The best way to learn to program is to do it.

2.1 Programming Skills Check

Take a look at: *https://data.gov.uk/dataset/road-accidents-safety-data*. This site provides public data on transport accidents in the UK. We will download this data and use it to find what roads has the most and fewest accidents in the UK.

Download and extract the 2015 Road Safety Accidents data. You can do this graphically or with the terminal commands,

```
$ wget http://data.dft.gov.uk/road-accidents-safety-data/ \
RoadSafetyData_Accidents_2015.zip[1]
$ unzip RoadSafetyData_Accidents_2015.zip
```

Take a quick look at the data with your own eyes. You can do this with the *less* command,

```
$ less Accidents_2015.csv
```

using space bar to step through one screen at a time and *q* to quit. (Or you can use a text editor to open it. As this is a large file, some graphical editors may run slowly, while *less* is designed to work well with large data files.)

Each line is one accident. The names of the data fields are shown in the top line.

---

[1]If this website is not available at the time of reading, a copy of the file is also provided in the Docker image, try *cd ~ /data/accidents*. *cd* means "change directory" and "~" is your home folder. There is also a solution program provided in the examples folder.

© Springer International Publishing AG 2018
C. Fox, *Data Science for Transport*, Springer Textbooks in Earth Sciences,
Geography and Environment, https://doi.org/10.1007/978-3-319-72953-4_2

Tasks:

- What weekdays have the most and fewest accidents?
- What are the mean and median number of vehicles involved in an accident?
- Plot a histogram showing accident distribution by time of day.
- How does the speed limit affect the number of casualties?
- What percentage of the UK's accidents happen in Derbyshire? Is this high or low relative to Derbyshire's population size?
- What other factors are important or interesting in relation to accidents?

Hints: Python has built in tools and libraries to split up lines of text into data fields, to maintain dictionaries, and perform sorting. There are some *fragments* of code below to provide hints for some of these operations.

```
#here is how to read data from a file:
for line in open("Accidents_2015.csv"):
    print(line)

#to extract data fields from lines of a file:
data_line = "here, is , some, data, 1, 2, 3 "  #a fake line
fields = data_line.split(",")
print("number of accidents in "+fields[0]+" is "+fields[2] )
#
#a small dictionary of Leeds road names and accident counts:
d=dict()
d["M62"]=0
d["M1"]=0
d["M62"] += 1   #add one accident
#
#don't worry about how these 2 lines work
#(they sort a dictionary of counts by value)
import operator       #this loads a sorting library
sorted_d = sorted(d.items(), key=operator.itemgetter(1))
print("the road with the FEWEST accidents is:" )
print(sorted_d[0])
print("the road with the MOST accidents is:")
print(sorted_d[-1])      #-1 is index of final element
#
#to convert between numbers as values and as text:
x=1
x_as_text = str(x)
x_as_integer = int(x_as_text)
#
#drawing a bar chart from a list
from matplotlib.pyplot import *
plot([4,7,6]); show()
```

2.2 Programming Languages

Data Science work can be done in most modern computer languages and your choice of language can be made (ideally) yourself by personal taste and (realistically) by your boss as whatever your organization is already using. All modern computer languages can compute the same things. They differ in two ways: first by the 'core' structures of the languages themselves, comprising internal syntax (the symbols and grammar of the language) and semantics (the meanings of the symbols); and second by the sets of libraries and tools that have been provided for them by the communities of their users.

At the time of writing, the most popular languages for working with data include Python and R. Both have quite similar syntax and semantics. Python is a general purpose language with many libraries and tools beyond Data Science, whilst R's community has focused solely on Data Science. R builds some Data Science features (such as data frames) into its syntax and semantics while Python provides similar features through libraries.[2] Python data programming thus typically requires a little more effort than R to set libraries up for doing Data Science, but then allows the programmer to work with and interface to many other types of system. It encourages the programmer to try out various competing libraries for Data Science rather than building any of them into its core. Like most computer systems, there have been several different versions of Python, and we will use Python 3 in this book. There are also various sub-versions, such as 3.6.2, at the time of writing, whose differences are not important for beginners.

Many of the libraries and tools used in Data Science are not written in either Python or R, but in others such as C, then have interfaces (or "wrappers") provided for many other languages including Python and R. These interfaces will look very similar regardless of the programming language used, so the learning of them via any language is generally transferable to use in any other.

In particular, core database functions are provided not by the programming language but by an external database program, which presents an interface to the programming language. Almost all modern databases use an additional language called SQL as this interface, which is covered in the next chapter.

A few specialist libraries are written in, and only usable from, Python or R themselves, such as some of the machine learning tools used later in this book. In general this is unfortunate and the world would be a better place if and when such libraries are usable, like their C counterparts, from arbitrary user languages rather than just the one in which they happen to be written.

2.3 Programming Environment

Data Science is difference from other forms of programming such as Software Engineering because it is inherently interactive. In Software Engineering, large "architectures" and "systems" are often designed on paper and white-boards then implemented and tested one component at a time before being integrated together. Doing Data Science is massively more enjoyable, interesting and efficient if we use the computer more interactively, giving commands and seeing the output straight away to guide our ideas and models.

IPython (Interactive Python) is a program that provides a nice interactive environment for use of the Python language in this way. It provides a command line which accepts typed commands, prints responses to them, and (with some extra libraries) launches other processes such as graphical plots. There are several ways to use IPython, but in this book we will use it via the Spyder interface launched

---

[2] For an introduction to R programming, see the excellent *Efficient R Programming: A Practical Guide to Smarter Programming* by Gillespie and Lovelace, O'Reilly, 2016.

in the previous chapter.[3] Sypder is an example of an Integrated Development Environment (IDE). An IDE is a graphical program that provides windows, icons and mouse control of both the IPython shell and its own editor. You can choose to give commands immediately to the IPython shell and/or to type sequences of commands (programs) into its text editor then run them together. Most large programs contain more than one file, and files refer to one another to link them into a single program made of many parts. To run a command, type it into the IPython window within Sypder. To run a program, create a file in Spyder's editor, save it, then run it by clicking the "run" icon on the menu bar (which shows a green triangle like a music "play" button).

Spyder has many tools and keyboard shortcuts which you can explore through its menus and manuals. We will mention just two particularly useful ones here: hold down *Shift-Control-I* and *Shift-Control-E* to move the cursor between the IPython shell and the editor – most users will do this thousands of times per day so these keys are worth learning straight away to save time using the mouse!

2.4 Core Language

It is traditional to begin learning any new programming language by printing (displaying) "hello world". This can be done from the IPython window of Spyder as,

```
print('hello world')
```

(and pressing return). We can also print arithmetic values such as,

```
print(2+2)
```

A *variable* is a symbol which can be given different values. For example,

```
x=4
print(x+1)
```

Python is a *typed* language which means that variables each have an "type" describing what kind of thing they are. Integers are a type, as in *x* above.

Real numbers (or rather, approximations called floating point numbers or *floats*) are another type and are denoted by including a decimal point such as,

```
y=2.5
```

or,

```
y=2.
```

Strings are the type of sequences of letters and other symbols ("characters") that form text such as "hello world", and are always denoted in (either single or double) quotes.

Operators are symbols which act on values to make new values, such as addition. The symbol "+" performs addition both on integers and floats. It is also used to join two strings such as,

```
s='hello'+' '+'world'
print(s)
```

---

[3] If you already like *vi*, *emacs*, or some other editor then you will probably want to stick with them. You can launch *ipython* directly from the command line then enter these editors using its *ed* command. Jupyter is another popular interface which runs inside a web browser and allows you to mix text and reports with graphics and code.

2.4 Core Language

To see the type of a variable, write,

```
type(x)
```

Sometimes we need to *cast* (convert) a variable from one type to another, such as,

```
x=1
print('There are '+str(x)+' cars')
```

which converts x into a string so that the '+' operator acts to string it together with the other texts, rather than try to add it numerically to them.

The basic string, integer, float, and other types can be quickly cast and inserted into strings using so-called "print-F" syntax,[4] which uses codes such as *i* for integer, *s* for string and *f* for float and it used heavily in data processing,

```
print('Camera %i saw car %s with confidence %f'%(5,'AB05 FDE',0.6))
```

Comments are notes for human readers which are ignored by the computer, and are prefixed with the hash symbol,

```
#this is a comment
```

2.4.1 Lists

A *list* is a type of variable which stores an ordered sequence of other variables, such as,

```
l=[1,2,'car', 5.0, [5,6] ]
```

Lists can contain variables of different types, including other lists like the final element above.
A single element can be accessed from the list like this,

```
print(l[4])
l[4]=9
```

A new list can be formed by selecting a subset of the elements from another, such as,

```
l[0:4]
```

These also work for strings,

```
s = 'hello world'
s[1]
s[0:4]
```

Note that Python counts ("indexes") the members of lists and other structures starting at 0 rather than 1, like British rather than American office buildings which have a zeroth "ground" floor with a "first" floor above them as the next in the sequence.

To add and remove elements from a list,

```
l.append('newstring')
l.remove(2)
```

---

[4]The name is a historical legacy from C-like languages, and the syntax is mostly compatible with those languages.

2.4.2 Dictionaries

Dictionaries (known as "hash tables" in some languages) are another basic type which generalize the above idea of a list. A list has several elements which each have an integer as their index. This allows elements to be located quickly and easily, for example by saying "give me element number 456", rather than having to search through all the previous 455 elements to find it. Imagine that a list is implemented as a series of physical pigeonhole mailboxes for an office. Each pigeonhole has a number as its label (its index) and stores things inside (its element). Now imagine that we replace these numeric labels with arbitrary variables. For example, we could use strings, then each element can be addressed by a name such as the following mapping from driver names to licence plates,

```
d=dict()
d['Andrew Froggatt']='XY32 AJF'
d['Andrew Bower']='XZ85 AJB'
```

From a Data Science view, dictionaries can be though of as a very simple form of "database" which allow for key-value pair storage. We will see much later that this type of store has become popular on "big data" scales through some of the recent "NoSQL" databases.

2.4.3 Control Structures

Like most programming languages, Python provides *loops* and *conditionals*, such as,

```
for i in range(0,10,4):
    print('hello world'+str(i))
x=1
while x<10:
    x=x+1
    print('while '+str(x))
    if x==1:
        print('x is one!')
    else:
        print('x is not one!')
```

Note that the indentations are critically important in Python, and maybe be written using either spaces or the tab character.[5]

2.4.4 Files

Before we can work with data or bring it into a database, we usually need to load it from files. Files are usually read and written one line at a time.

Here is how to write a file,

---

[5] Debate over which is best is still raging. The present author supports tabs because a single tab is more minimal than multiple spaces!

```
f=open('myfilename.txt', 'w')
f.write('hello')
f.close()
```

Here is how to read a file,

```
for line in open('myfilename.txt'):
    print(line)
```

2.4.5 Functions

A function is a self-contained piece of code which takes some inputs and returns an output, and can called many times from other parts of a program. Functions are defined and used in Python like this,

```
def myfunction(a,b):    # a,b are the inputs
    r = a+b
    return r            # return an output
x=1
y=2
z=myfunction(x,y)       # call the function
print(z)
```

Python can be used for object-oriented programming (OOP) which allows for complex data types ("classes") to be defined along with functions that operate on them. We will not make much use of OOP in this book but will sometimes encounter objects ("instances" of classes) that have been created for us by libraries and tools. When this occurs, the name of the function is appended to the name of the object by a dot, such as,

```
myobject.myfunction(a,b)
```

which is roughly equivalent to writing,

```
myfunction(myobject, a,b).
```

2.5 Libraries

Most of Python's Data Science functions are provided by popular libraries rather than the core language. It is not possible to describe all of them here, but see the further reading section for full references. Usually if you feel that some function should exist in a library, then it probably does.[6]

2.5.1 Modules

When programs start to get big – typically when they fill more than one screen of text – it usually makes sense to split them into several smaller files, which are known as *modules* in Python (or "libraries"

---

[6]If it really doesn't, then you can add it to the library yourself and publish it for others to use – all these tools are open-source.

or "components" in other languages). As well as making it easier to navigate through the code, this has the advantage that modules can be reused by multiple other programs. Python modules are saved with a *.py* suffix. You can reuse a module in a new program like this,

File *mymodule.py*,

```
def myfunction(a,b):

    return a+b
```

File *myprogram.py*,

```
import mymodule
z = mymodule.myfunction(1,2)
```

An alternative syntax for the reusing program is,

```
from mymodule import *
z = myfunction(1,2)
```

Libraries for Python always come as modules. The Docker image supplied with this book has many set up ready to use. You will see them appear as import statements at the start of programs throughout the book.[7]

2.5.2 Mathematics

Python's *math* library provides basic mathematics operations such as,

```
import math
print( math.sin(math.pi * 2) )
print( math.exp(2) )
```

(Unlike some languages, the power function is part of the core language and written *10**2* for 10 to the power of 2.)

Vectors, arrays and matrices (linear algebra) are provided by the *Numpy* library, which by long standing convention is renamed as *np* to be more concise,

```
import numpy as np
Z = np.zeros((2,3))
I = np.eye(3)
A = np.matrix([[1,2],[3,4]])
A[0,1] = 6
print(A[0:2, 1])
print(A.shape)
print(A.dot(A))  #matrix multiplication
print(A+1)       #add scalar
```

---

[7] There are several ways to install new Python libraries on your computer, the most common at the time of writing is a tool called *pip3*.

2.5.3 Plotting

Plotting graphs in Python is usually done with the *Matplotlib* library like this,

```
from matplotlib.pyplot import *
xs = [1,2,3]
ys = [10, 12, 11]
plot(xs, ys, 'bx') #blue x's
hold(True) #add future plots to the same figure
plot(xs, ys, 'r-', linewidth=5) #thick red lines
text(xs[1],ys[1],'some text')
title('my graph')
ylabel('vehicle count')
gca().invert_yaxis() #flip the y axis
xticks( range(0,16), ['car', 'van', 'truck'], rotation='vertical')
```

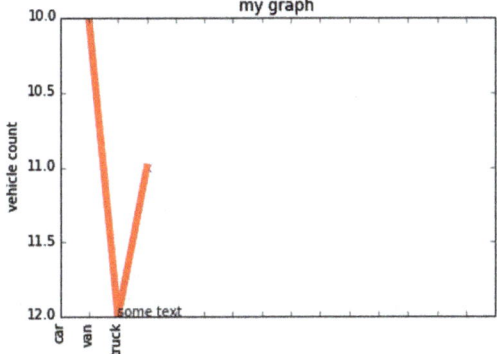

It is possible to manipulate plots in most ways that you can imagine or need if you consult the Matplotlib reference documents. Plotting lines with different widths is especially useful to show flows in transport networks.

When plotting from within Spyder, the default setting is for Spyder to display the plot inside the IPython window, known as "inline" mode. This is usually what you want for Data Science work but sometimes you will prefer to have it appear in a new window of its own (in particular, this allows you to pan and zoom it with the mouse). This mode can be selected using Spyder's menus: *Tools → Preferences → IPython → Graphics → Backend → Automatic*.

2.5.4 Data Frames

Data frames are data structures similar to LibreOffice Calc or Microsoft Excel spreadsheets, or to database tables, having data in multiple rows of named columns that each store different types of data such as numbers and text. They are available in most modern languages including Python, via its *Pandas* library (often renamed as *pd*). We will see later that Pandas is able to interface with large databases, but more basically it can also read and write data frames (*df*) from CSV files. This saves time and effort on manually reading such files with loops and breaking down their data to store in arrays. Some useful Pandas commands are given below for reference. Some may not make sense on first reading but will become a useful reference for exercises later on.

```
import pandas as pd
import numpy as np
df=pd.read_csv('data/accidents/Accidents_2015.csv') #load from csv file
```

```
df.columns                                           #names of columns
df.shape                                             #number of rows and cols
df['Number_of_Vehicles']                             #extract a column
df[0:20]                                             #extract some rows
df.iloc[7]                                           #extract single row
df.iloc[::10, :]                                     #extract every 10th row
df.as_matrix()                                       #convert to numpy
I = np.eye(3)                                        #create a matrix
df = pd.DataFrame(I, columns=['col1','col2','col3']) #convert np to pd
df['col1']+df['col2']                                #add columns
df + df                                              #add all columns
del df['col1']                                       #delete a column
df.append(df)                                        #append frames
df.sort_values('col2', ascending=False)              #sort by a column
df['newField']=0.                                    #add extra column
df[df['col2']==0]                                    #select rows WHERE true
df.merge(df)                                         #JOIN columns from frames
df = df1.merge(df, how='outer').fillna(method='ffill') #align by time
df.to_csv('filename')                                #save frame as CSV
```

2.5.5 Debugging

Debuggers allow programmers to make the program stop at a predetermined line in the code where they can manually inspect values of all variables interactively, including drawing plots and performing computations on the values. Spyder includes a graphical debugger. To use it, double click a line of code in the editor to set a breakpoint (appears as a red circle). Press *Control-F5* to run in debug mode, which will stop at the breakpoint. Use the IPython shell to inspect variables, and the blue arrows in the menu bar to advance through the program one line at a time or in and out of functions.

Many programmers spend the bulk of their working days using debuggers whilst others consider reliance on them a sign of poor software design or prefer to track errors by printing out values of many variables from within the code itself.[8]

Data Science is different from other forms of software engineering in that many problems are caused not by program bugs but by unexpected "glitches" in the data. Typically the programmer makes some reasonable-seeming assumption about the data, such as values in some field being above zero, then their program dies when an unexpected negative number is encountered. Glitches can take a long time to track down, are (arguably) not the fault of the programmer, and often require interactive plotting of data at the failure point, so heavy debugger use is perhaps more widely accepted in this context than other forms of software engineering.

2.6 Further Reading

The official Python tutorials, as used by the vast majority of new (and old) Python programmers,

- https://docs.python.org/3/tutorial/index.html

---

[8] More rigorous approaches to reducing errors are researched by the *functional programming* movement, which emphasises programs as defining mathematical function rather than how to compute them step-by-step, often including heavy use of type systems as a means to catch errors; and the related *formal methods* movement which researches ways to mathematically prove programs to be correct. Functional programming has gained in popularity in recent years, while formal methods remain a research area in most cases.

2.6 Further Reading

Numpy tutorial,

- https://docs.scipy.org/doc/numpy-dev/user/quickstart.html

Numpy for Octave/Matlab users will be very familiar, so much so that many commands will run directly or with a few changes in Numpy. (Just remember to index from 0 instead of 1.) A detailed conversion guide for Octave/Matlab users is here,

- https://docs.scipy.org/doc/numpy-dev/user/numpy-for-matlab-users.html

Pandas usage,

- http://pandas.pydata.org/pandas-docs/stable/tutorials.html
- McKinney W (2017) Python for data analysis, 2nd edn. O'Reilly

Matplotlib usage,

- http://jakevdp.github.io/mpl_tutorial/tutorial_pages/tut1.html

Python IDEs, debuggers, testing, and related tools,

- Rother K (2017) Pro python best practices: debugging, testing and maintenance. Apress

Database Design 3

3.1 Before the Relational Model

During the Python exercise, we used text processing operations to step through a CSV file and process each line at a time. For some applications, this method is scaled up to store and process larger data sets. For example we might have a separate CSV file for each year's road accidents for many years, and perhaps also for many countries.

We often store small sets of files on our home computers in this way, using files and directories. A problem that we often run into with this structure is that the directory hierarchy imposes an ordering on how the files are stored by their type. For example, for accident lists per country and per year, we might choose to use directory names like */accidents/UK/2015* and */accidents/USA/2016*; or alternatively we could use names like */accidents/2015/UK.csv* and */accidents/2016/USA.csv*. When we write programs to process this data, they will then be structured to loop over the directories such as,

```
for countryID in os.listdir("accidents/"):
  for yearID in os.listdir("accidents/"+countryID+"/"):
    for line in open("accidents/"+countryID+"/"+yearID):
      print(line)
```

For tasks requiring iteration over every accident in the set, this is sufficient to work. But what if we want to iterate over just accidents from one country or from one year? A naïve method would be to write a program like this,

```
for countryID in os.listdir("accidents/"):
  for yearID in os.listdir("accidents/"+countryID+"/"):
    if countryID=="UK":
      for line in open("accidents/"+countryID+"/"+yearID):
        print(line)
```

The problem with this approach is that the program is still performing the computational work of iterating over the entire database every time a task is performed, even if the task only requires a small subset of the data. In some cases, we can optimize the data structure for particular tasks. For example, if we know that we will often want to work with all the data from one country but not (or less often) from one year, then we could choose the storage convention "*/accidents/UK/2015*", and write programs like,

```
countryID ="UK"
for yearID in os.listdir("accidents/"+countryID+"/"):
  if countryID=="UK":
    for line in open("accidents/"+countryID+"/"+yearID):
      print(line)
```

This is a very brittle solution though, because usually in real life, your manager will eventually decide to ask you to start making computations that select by year or some other variable. The decision you made at the start about the directory structure will then come back to haunt you, and will be very hard to change because all your code is now written assuming that format of file names.

Many people run into this problem quite quickly when then try to organize their music collections – either in physical or digital form. Ordering a set of CD cases on a shelf can be done by the name of the composer, performer, or album title. There are usually some discs containing work by "various artists" that cause particular problems.

A related problem with the directory-file model is with the files themselves. If the lists of accidents grows very large (such as 4 MB in our example), then even once we have located the right file, we might still have to spend a large amount of search time looking for particular lines within the file. For example, once we have the accidents for the UK in 2015, we might want to find all those that are in Chesterfield, by searching though the whole file and rejecting lines that are not in Chesterfield. We could move this information out of the file itself and into the directory structure, for example by making new sub-directories with names like */accidents/UK/2015/Chesterfield.csv* and */accidents/UK/2015/Leeds.csv*. However this would again enforce a particular ordering on these fields. (Conversely, we could move all the information from all the files into one big file, with new columns for country and year, which would make searching within the file equally poor.)

You might have tried a solution to your computer music organization problem that makes use of "soft-links" or "virtual directories" in the file hierarchy. This is one way to try to solve the hierarchy problem because it allows you to have multiple paths such as */accidents/UK/2015/Chesterfield* and */accidents/2015/UK/Chesterfield* pointing to the same file. This is called a "network model" and can sometimes help a bit. But unless you create an exponential number of virtual directories for every possible ordering of the variables, there will still be cases that it doesn't cover.

Hierarchical directory and file structures are still in common use as a form of database in many Data Science applications. For example, pedestrian detection from the EU *CityMobil2* self-driving car project is stored as files like */cm2/experimentDate/ vehicleNumber/experimentNumber/detectionNumber.csv* and is sufficient for many purposes.

3.2 Picturing the World

The *relational model* was developed in 1970 to address the file hierarchy problem, and remained the main database model until the present decade. In recent years it has been challenged by other "big data" approaches, but it remains a large influence on most new methods. It is likely that most transport offices will currently be using relational data, even if they are currently thinking about "big data" methods too. The core idea of the relational model, and of many other models, is to separate the implementation of the data storage from the commands used to express operations on it. In Computer Science this process is known as *encapsulation*, and the set of commands is an "Application Programmer Interface" (API). Encapsulation is one of Computer Science's main tools for dealing with and reducing complexity of software. Instead of writing programs to operate on raw files, as an analyst you will write programs that operate on your "picture of the world" as described by the data. The database program – together with some implementation decisions made by the database architect – is then responsible for figuring out how to make these operations run in efficient ways. You don't need to care how the data is stored in files, or (ideally) on how many hard discs or with how many computers used to process it. You just ask the system to operate on the data and it gives you the result. A database appears to its users as a picture of the world rather than as a computational structure, and its job is to represent the world. What kind of pictures should we use in this representation?

3.2.1 Ontology

The encapsulation idea is very general and covers many models of data. To understand what a relational database is, and how it differs from other models, we need to consider ontology. We need to talk about *things*.

In Philosophy, Ontology is the study of "things". In ancient times, this was considered as a debate over what "really" exists, with some people arguing that the world was made of water, or four elements, or atoms. In modern philosophy, it is generally understood that humans can never know what "really" exists at all, but instead that different people and groups of people maintain their own mental models to describe patterns in their experiences, which are useful for different purposes. Ontology then becomes the study of what these models are, or should be, for various people and purposes.

"Ontology" is used in two slightly different senses. In its purest form, which we will call "Philosophical Ontology", it asks what the general structure of mental "things" is. This is the question that relates to relational, object-oriented, and other data models. Once a decision is made at this high level, such as to use a relational model, then the secondary form, which we will call "Data Ontology" (but is perhaps more similar to "Metaphysics" in Philosophy), asks how to model some specific set of "things" for a particular purpose within a philosophical ontology. In this case we can speak of "an ontology" or "some ontologies" used to model different data sets within one Philosophical Ontology paradigm.

Philosophical Ontology is usually missing from popular and professional database texts, which begin with Data Ontology and implicitly assume the relational model for the philosophy. The relational model has held up for almost 50 years as the mainstay of practical databases, however in the last decade we have begun to see it being questioned due to the computational needs of "big data" conflicting with some of its requirements. If it does have to change then data scientists will need to think about Philosophical Ontology again as they consider replacement for the relational model.

3.2.2 Philosophical Ontology

Philosophical Ontology is a classical branch of philosophy, sometimes considered to be its purest and most abstract branch. It asks questions about the basic categories of being or existence. It is often charactatured (quite accurately!) as two Oxford professors sitting in armchairs, debating questions like,

"Does the redness of apples belong to the apples?"
"Is the road from Athens to Thebes the same as the road from Thebes to Athens?"
"What's red and invisible?" (Answer: "no apples.")

To most engineers these questions seem like a waste of time, and it is easy to mock philosophers (Fig. 3.1) for apparently continuing to disagree over their answers after trying for 2,500 years!

However, as data scientists we do need to take them seriously, because they become very practical design issues when we ask how a database encapsulation should represent the world. What components

Fig. 3.1 From *www.existentialccomics.com*.

3.2 Picturing the World

are our mental models of the world made of, and how should we represent them in the interface to our databases?

When we ask what are the "things" that the (mental) world is made of, we do not mean what *particular* things it is made of, such as cars or accidents or atoms. Rather, we ask what do we mean by a "thing" in the first place, such that cars and accidents and atoms and colours can all be types of these "things". Different philosophers have proposed different definitions of "things" which have led directly to different models of databases and programming.

- How would you define a "thing"?
- What makes two "things" different from each other?

Most normal people will initially give an answer something like:

> "A *thing* is an entity which has a position in space, and some properties, like colour or size or weight."

This is a good starting point for database design, and is roughly the philosophical ontology expressed by the classical philosopher Aristotle: we might model the world by a set of entities (or "objects") which each have a position and some set of properties. The connection of an entity to its property is called a *has-a* relationship. If we look carefully we see that this definition actually describes three different kinds of "thing": entities, locations and properties. We might consider position simply as another property, so that there are two types of thing – entities and properties – where entities may be modelled by a line in a CSV file, with commas separating values of a standard set of properties, such as,

entity	position	colour	size
myCar	Woodhouse Lane carpark	silver	medium
ANPR_cam57	M1 J36 bridge south facing	yellow	small

- What are the limits of this idea of a "thing"?

Clearly this idea of a "thing" is rather limited in several ways. For example it doesn't represent the speed that my car is moving, or the car's age, or its licence plate number. However we cannot just add these properties to the table of things because most of them are not applicable to the ANPR camera. ANPR cameras have different properties from cars, such as frame rates, detection accuracies, and hashing functions. To model this, we need a better idea of what "entities" are, which includes a "type" or "class". Such as,

> "An entity is a member of a class, and has values instantiating all the properties of that class".

Now we have three types of "thing" in the world: entities, properties, and classes. To make this work we also need a definition of what a class is, such as,

> "A class is a set of properties".

Most databases and many programming languages tighten this to "a class is a set of properties with specified types", so that weight must be a number of kilograms and age must be a number of years etc. We still need a definition of "type", which is usually a mix of raw values such as "integer", "string", "real number" and recursion, allowing existing classes as types in new class definitions. Under this view of the world, each class can be modelled as a separate table, containing a list of entities of that class and their properties, such as,

Cars:	position	speed	age
myCar	Woodhouse Lane carpark	0	5
richardsCar	M62 J4	70	4

Cameras:	position	frame rate	detection rate
ANPR_cam57	M1 J36 bridge south facing	10	0.99
ANPR_cam59	M1 J365 bridge north facing	20	0.95

This works well for many static databases and computer programs, and is the foundation of "object-oriented programming" as used in Python's class system. Despite this, philosophers continue to debate this model. The famous "problem of universals" asks, "in what sense do properties exist if there are no entities instantiating them? (hence the deeply profound "no apples" joke above). Similarly, Plato is usually understood to have believed in the independent existence of classes ("forms") without entity instantiations.

The basic noun/adjective structure of most natural languages mirrors this conception. But natural languages also contain verbs, which are so far missing from it. Especially in transport, which studies moving things, we need a way to express notions of time and change. Is the *mycar* the same thing on Monday as one Tuesday or is it a different thing then? What if the owner has changed its engine, or its colour, or its licence plate? Is the M1 motorway the same thing on Monday and Tuesday? It may have similar traffic flows but they will be made up of different cars. Lanes may have opened or closed, or even been added by construction work. Can you cross the same motorway twice?

- How might you represent *mycar* moving along a motorway?

One way to represent time and change is to consider an *event* as a higher level thing than an entity. We could consider that, like a class, an entity is not directly observable. Rather, we observe events which are each comprised of "an entity at a time", such as,

Events:	entity	date	time
accident1	mycar	2017-01-20	07:30
detection11	ANPR_cam57	2017-01-15	15:37

We have previously considered location to be a property of an entity, while its time is a property of an event rather than a thing. This gives different status to space and time. For applications like transport this is usually OK, even though modern Physics prefers to treat space and time more equally. In 20th century "process philosophy", attempts have been made to conceive of events as primary rather than things, partly in response to changes in Physics and partly to enable clearer thinking about everyday events. This remains a research area in Philosophy and might one day provide inspiration for new database ideas.

Is this now a complete theory of Ontology? No – we usually also need to conceive of interactions or relationships between two or more things in order to say anything useful about the world. These are usually called *relations*. For example "the cat is on the mat" is a spatial relation between two things. "*car1* overtakes *car2*" is a relation in both space and time. In natural English, some relations are indicated by prepositions such as "on" and "next to", and other by verbs.

As with raw properties, timeless entities and abstract classes, philosophers may have fun debating the existence of relations. Is the overtaking of *car2* by *car1* a property of *car1*, or of *car2*, or of both, or neither? Is it something other than a property? If so – then the world is now made of five kinds of things: entities, properties, classes, events and relations. Could a relation be just an entity? It must have its own properties which connect it to *car1* and to *car2*. If it has properties, then what makes it different from any other kind of entity? But somehow we feel that *car1* and *car2* are more fundamental or more "real" than this overtaking relation between them. (Surely everyone will see the two cars as separate entities, but perhaps only people who know about driving will see the overtaking?) Alternatively we might remove events from our ontology by considering them to be a just special case of relations between entities and times.

The object-oriented programming model tends to have problems with events and relations. It is very good at modeling entities, classes and properties. But events and relations are usually fudged by modeling them as new classes containing entities and times. An object-oriented model will likely have classes with names like *Car*, *CarAtTime*, and *OvertakingEvent*. A problem (arguably) with this approach is that it fixes the space of possible event and relation types in advance. If we later become interested in an event which is a particular human driver overtaking, rather than their car, we have to go back to the software designers and ask them to make a new *PersonOvertakingEvent* to be added to the program. We can't conceive of that new event type on the fly during our data analysis. Worse, we get the same hierarchy problem as with using text files in directories as models: a class modeling a relation between two events is different from a class modeling an event containing two relations; and a relation over time between two spatial events is different from a relation over space between two temporal ones.

The relational model is more general than object-oriented programming because it allows for such new types of event and relation to be composed on the fly, and works with the entire space of possible combinations rather than a predefined set of classes. This may be considered good thing or a bad thing. It is a good thing because it avoids the hierarchy problem and the predefined classes problem. It is a bad thing because most programming languages assume the object-oriented model, which leads to clashes between ontologies when a program interacts with a database. The relational model simplifies the collection of objects, events and relations into a single notion of a typed relation, having a class.

The "modernist" philosopher Wittgenstein (who also worked as an engineer, and would probably have been a database designer if alive today), began his philosophy with,

1 The world is all that is the case.
1.1 The world is the totality of facts, not of things.
1.11 The world is determined by the facts, and by their being all the facts.
1.12 For the totality of facts determines what is the case, and also whatever is not the case.
1.13 The facts in logical space are the world.
1.2 The world divides into facts.
1.21 Each item can be the case or not the case while everything else remains the same.

This is essentially the philosophy which is enshrined in the relational model. Rather than having entities, events, and relations, we only have typed relations (Wittgenstein calls them "facts"). A typed relation is an instance of an ordered, typed, *n*-tuple; from a class defining the size, ordering, and types. For example we can represent *Car* entities with typed-relations from the class *Car(name, position, speed, age)* such as *Car(MyCar, WoodHouseLane, 0, 5)*. We can represent events and relations with typed relations such as *Accident(car1, car2, time)*. Unlike in object-oriented programming, relations may be operated on to give relations of new classes on the fly. For example if we ask,

"Select all ages and times of cars in accidents where the accident time was after 3pm.",

then we can define the result's type as a relation of the class *NewRelation(age, time)*. This type does not correspond to the types of either *Car* or *Accident*, it is a new type of relation that arises only when this question is asked.

This is the view proposed by (Codd, 1970) which remains the foundation of relational databases and of most deployed databases online today. As well as the relational model's tensions with object-oriented programming, it is possible that the computational needs of "big data" will force a move back to object-oriented styles or even to file based models. This may need some careful thought about how their ontologies will change if we are to avoid their historical problems.

- Do you think the world is made of entities, or made of facts?
- Which do you think is the better way to model it for Data Science?

3.2.3 Data Ontology

The issues in Philosophical Ontology are very abstract compared with the daily work of a database designer, which typically takes either the relational or object-oriented model as given, then proceeds to worry about more practical Data Ontology. Assuming the relational model, Data Ontology consists of choosing particular classes of typed relations to model and solve particular problems. This is a "neat" or "modernist", data process which is specific to each problem. Modernist database designers are very careful with their designs, avoiding contradictions and representing consistent worlds. (They can thus be very protective about "their" databases, especially from new staff wanting to insert their own data or models into them!) This "modernist" view is currently under pressure from "messy" or "post-modern" big-data researchers who focus more more re-using old data for new, unintended purposes. But we describe only the modernist approach for now.

The modernist design process starts with a client and a problem. The client is usually a specialist in a particular domain, such as transport. The client wants to know the answer to some question which is formulated in terms of everyday "things" such as cars, speeds, and accidents. The database designer works with the client to understand what the entities are, and which of their properties are important. It is not (in modernism) desirable or useful to model entities and properties that are irrelevant to the client's problem. For example, if the task is origin-destination analysis then we do not model the colours of cars.

Sometimes, *sub-classes* will be used to describe different but related kinds of entity. For example, all *Vehicle*s may have a weight, licence plate and speed, but *Truck*s are a sub-class of *Vehicle* which also have a cargo type; while *Car*s may have a *numberOfPassengers* and *numberOfDoors*. Sub-classing may be modelled with an *is-a* relation, which says that one entity is-a subclass of another.

In natural English[1], we use the word "is" to mean (arguably) four different things: existence (there is a car), *has-a* (the car is red), *is-a* (the car is a vehicle), and *is-related* (the car is on the road). With

---

[1] The dual-meaning of the existential "is" with "has-a" is shared by most Indo-European languages descended from ancient Persian. In some languages these are represented differently, for example one might "have" a property rather than "be" it. For example, "I have redness" vs "I am red". Occasionally this appears in English, for example "I have 20 years of experience" is similar to the French "J'ai vingt ans".

3.2 Picturing the World

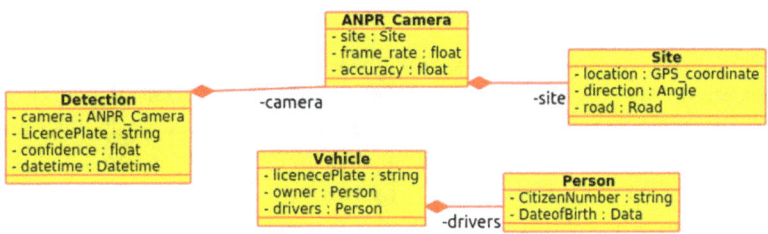

Fig. 3.2 UML model for an ANPR database

data we have to be more clear. Some of these types of "is-ness" or "being" will be modelled in data by properties, and others in other ways. For example, in some object-oriented programming styles (such as doing OOP in straight C), *is-a* is represented as a property; while in other styles (such as C++) it is given its own special syntax and structures.[2]

The graphical *Unified Modeling Language (UML)* is often used, both formally and informally, during the design process. UML shows each class as a box, containing slots for properties. Some of the properties are names of entities from other classes, which are shown by arrows pointing out to them. The *is-a* relationship is shown by another type of arrow. Figure 3.2 shows an example of a small UML database design for modeling ANPR detections.

- What are some problems and limitations of this data ontology?

Even with a simple model like this, we already find many ontological problems. Here we have modelled *Camera*s, *Detection*, *Site*s, *Vehicle*s and *Person*s as relations, while other entities such as *licencePlate*s and *date*s are considered as properties. Why should a *Camera* have a different ontological status from a *licencePlate*'s? We could alternately have modelled *LicencePlate* as an entity containing a string containing is letters and numbers. This would make the UML look prettier because then the *licencePlate* properties of the *Camera* and of *Vehicle* would have something to connect to. It is not obvious that they refer to the same thing otherwise. Is class *Vehicle* sufficient for our model, or should we have subclasses *Car* and *Truck* to enable questions to be asked about those entities? Or is *vehicleType* a property of a *Vehicle* rather than a sub-classing relationship? Is the confidence of a *Detection* a property of that detection, or is it a higher level property about a belief that a particular detection did or did not exist? What is going to happen when a vehicle has two or more registered drivers, and how do we model which one was present at the detection event?

There are a few basic principles of data ontology design which are either computationally or practically useful and should usually be followed:

3.2.3.1 No Redundancy

Each piece of information should be represented once and only once in the database. For example, we might be tempted to have properties like *DriverName* and *DriverDataOfBirth* in the *Detection* event. Several properties like this appear in the *Accidents.csv* file seen previously, to make them easy to represent in a single file. The problem with this is that this information may then appear in multiple places, for example also in the *Person* class. This makes it possible for inconsistencies to arise in the database. If there is any possibility that data contained in a class will be reused by other classes, then consider pulling out into its own class.

---

[2]See the constructed language "Lojban" for an example of something in-between natural and data languages, using explicitly logical relations. No-one actually speaks Lojban but studying it is a great way for natural and data languages and ontologies to learn from each other.

3.2.3.2 Sense and Reference

Where an entity has a property that is a member of another class (e.g. a *Camera* has a *Site*) then the *Camera* should store only the *name* of the Site, rather than the *Site* itself. (This is not so clear in Codd's paper, which might be read as suggesting that relations actually contain other relations as properties, rather than just their names.) This means that each *Site* must be given a unique identifier, such as a number or codename, called a *key*. We say that the key is a "sense" of the *Camera*'s property, and the *Site* data which it refers to its its "reference". Another way to specify an entity's sense is to describe some of its properties, for example "the car detected at 9:07" or "the car detected at 19:36".

In the relational model, name keys appear as just another property and have no difference in ontological status from other properties that describe real-world data properties. This can sometimes lead to problems where the same real-world entity appears in the database under two different names. For example, our camera might detect a Morning Car and and Evening Car during a day, with their licence plates obscured. It might be the case that the two entities created, *MorningCar* and *EveningCar*, refer in reality to the same vehicle, but the system does not know this. Perhaps at a later date we get a clearer picture of the licence plates and realize that both senses refer to the same entity, in which case we will have to merge them in the database. This will include reviewing all other classes that have either sense as a property and updating their values too.

Another use for keys is computational. We will often want to search quickly for entities having particular names. Most commonly, this happens when we want to pull out data from a property member of one class, for example, looking up details of a particular *Site* that corresponds to a known *Camera*. We will talk more about how database systems do this later in the book.

Try to use standard naming conventions for entities where possible. This will help if you later have to import data from other places, as there is a good chance that the designers of the other database will have used the same names. For example, vehicles can be named by their *licencePlate*, and people by their government citizen numbers.

3.2.3.3 Event Design

Although the relational model itself makes no distinction between entities, events and relations, it is usually useful to do so yourself. Design the entities first, then events and relations that link entities to places, times and to each other.

3.2.3.4 Use Case Modeling

Once you have a first draft design, work with your client to understand the types of questions they will want to ask of the data. Produce specific example questions to cover as many use cases as you can. Model your database design using paper and pens, and work through how the use cases will operate on your data. Update your designs when problems occur. It is useful to consider questions that the client might ask in the future. Often they do not yet know what their questions are, but a good design will anticipate some of them so that large data restructures are not needed in response to them. Common requests include breakdowns by city, or vehicle type, road type, road conditions, and date so you often want to include many of these from the start.

3.2.3.5 Data Normalization

Normalization is a formal process which includes both redundancy removal and also limits potential future redundancies as new classes are added. It was described in the original (Codd, 1970) paper and has remained important ever since. It begins by forming a class hierarchy tree, with the most abstract classes at the top. It works down through the tree, checking what properties can be pushed down the hierarchy from each class. This process is guaranteed to create a nice, clean database, where all entities are easily available for reuse by new classes and contain no redundancy.

3.2.4 Structured Query Language (SQL)

The relational model of (Codd, 1970) is implemented by the *Structured Query Language* (*SQL*, sometimes pronounced "sequel"). SQL arose when different database software companies wanted to make their systems interact with each other through a single standard interface. So in theory, you can take your SQL code written on one database and run it on a different database made by another organization. Codd defined needs for two separate languages: one for creating and manipulating the data ontology of a database, and another for working within that data ontology to insert and query specific data items. SQL has largely achieved its inter-interoperability goal for the second of these. But different database software using their own variations of the former remains a problem. Major SQL implementations include PostgreSQL used here; also MySQL, Oracle, and Microsoft SQL Server. PostgreSQL and MySQL are open source software which can be downloaded and run on most types of computer.

3.3 Exercises

3.3.1 Setting up PostgreSQL

Using *itsleeds,* start the database service running in the background of your computer: create a user; create a database; and log into it,

```
$ /etc/init.d/postgresql status #check if running
$ service postgresql start #start if it isn't already running
$ sudo -u postgres createuser -s root #create a user
$ sudo -u postgres createdb mydatabasename -O root #create a database
$ psql -d mydatabasename  #log into the database (gives SQL command line)
```

3.3.2 SQL Creation Language

PostgreSQL, like most database software, comes with a command line interface which lets you execute SQL commands. Many databases also have graphical tools (e.g. *pgadmin*) to automate aspects of SQL command writing, but it is best to learn to work with pure SQL first before using them, because with pure SQL you can see exactly what is going on at all times. Here is an example of creating tables for the *ANPR_Camera* and *Detection* entities, which specifies the types of their properties. These commands are typed at the *psql* prompt after launching it as above,

```
CREATE TABLE Detection (
id serial PRIMARY KEY,
camera integer NOT NULL,
licencePlate text,
confidence float,
timestamp timestamp
);
CREATE TABLE ANPR_Camera (
id serial PRIMARY KEY,
site integer NOT NULL,
```

```
frame_rate float,
accuracy float
);
```

By tradition, table names are *singular* names of the relation types, such as "Detection" rather than the plural of the data contained within them "Detections".[3] It is useful to stick to traditions because if everyone does so, then they will make fewer mistakes when working with each others' databases.

Some useful commands you can give to check your tables have created properly are,

```
\dt #describe tables currently in the database
\d+ Detection #describe properties of a particular table
```

(The results are displayed using *less*, so press *q* to quit or *Space* to see more.) Raw types include *text* which means text string (variable length characters); *timestamp* stores a date and time together; *checkbox* (multiple choice) data; *float* (floating point – real number approximation) and *serial*. *serial* is a special integer type which is automatically filled in with unique numbers for each new item, starting at 1 and incrementing. It is usually used, including here, to name entities. *PRIMARY KEY* is a special command which tells the database that this property will be used as a unique name.

To destroy a table (e.g. when you make a mistake and want to create it again), use the *DROP* command. Take care though as this will also delete any data that is stored in the table,

```
DROP TABLE IF EXISTS Detection;
```

3.3.3 SQL Query Language

To insert data into the database, after creating the tables, we use *INSERT* commands like this,

```
INSERT INTO Detection
(camera, licencePlate, confidence, timestamp)
VALUES (4, 'A01 NPR', 0.78, '2014-04-28:09:05:00');
INSERT INTO ANPR_Camera
(site, frame_rate, accuracy)
VALUES (76, 50, 0.99);
```

Note that when a table has a special serial property for naming, we don't insert that value because it will be filled automatically.

When we import data in bulk, it is common (and fast) to insert many rows of data in a single query rather than with several small queries. Such as,

```
INSERT INTO Detection
(camera, licencePlate, confidence, timestamp)
VALUES
(3, 'A02 NPR', 0.53, '2014-04-28:09:12:14'),
(4, 'A04 NPR', 0.28, '2014-04-28:09:17:35');
```

If you need to many multiple inserts, they will run faster if committed all together with a single commit than with individual commits.

To retrieve data from the database, we use the *SELECT* command. This is usually the most common command given to databases. To select everything ("*" means "all") from the database,

```
SELECT * FROM Detection;
```

---

[3] For object-oriented programmers: this is similar to naming classes such as "Detection", for the class of all detections.

3.3 Exercises

Often your tables are very large and you just want to view a sample just as the top 10 values, sorting by a property such as timestamp,

```
SELECT * FROM Detection ORDER BY timestamp LIMIT 2;
```

SQL's real power comes from queries that retrieve specific pieces of information. For example, to select for a value of a property,

```
SELECT * FROM Detection WHERE confidence>0.6;
```

Or to delete specific selection,

```
DELETE FROM Detection WHERE licencePlate='A01 NPR';
```

Following Codd's model, we can form new relations by restricting and joining the domains of existing relations. Restriction is easy, we just ask for a subset of properties rather than "all". (If you are used to object-oriented programming this may appear to be a strange concept, because the result is a relation with no clear "class".),

```
SELECT (camera, licencePlate)
FROM Detection
WHERE confidence > 0.5;
```

Joins are queries that combine information from multiple tables. Good design practice is to make lots of small tables, which push information about entities as far down the hierarchy as possible. To recover that information you need to join. Joining is computationally complex, but because of encapsulation we don't have to care about computation here, we just ask the database to do it for us! Joining makes use of entity names, which are often unique numerical IDs, such as *ANPR_Camera*'s *id* property here. We can retrieve data from a new, larger, relation by joining these tables together with an outer product, then restricting the set of all combinations to the ones we are interested in like this,

```
SELECT * FROM Detection
LEFT OUTER JOIN ANPR_camera
ON (Detection.camera = ANPR_camera.id);
```

(For more examples see *https://www.postgresql.org/docs/8.3/static/tutorial-join.html*).

SQL can compute and return aggregate values such as totals and averages, for different groups of data. For example the following will compute the mean confidence of all the detections for each camera,

```
SELECT camera, AVG(confidence) AS mean_confidence
FROM Detection
GROUP BY camera;
```

In particular, a common idiom is to ask for the total number of rows in a table, to measure how much data we have (where *COUNT* is an aggregator counting the number of entries),

```
SELECT COUNT(*) from Detection;
```

SQL is a complex language and has many more operations and ways of combining them, however many useful programs use just the above subset of commands. SQL is hierarchical so you can write things like,

```
SELECT * FROM (SELECT * FROM Detection
LEFT OUTER JOIN ANPR_camera
ON (Detection.camera = ANPR_camera.id)) AS foo
WHERE confidence>0.5 AND camera=3;
```

Hierarchical SQL queries like this can run very fast, even when they are very large, and will usually be faster than trying to achieve the same effect using several smaller queries.

3.3.4 SQL Python Binding

The SQL command line is useful for testing and inspecting databases, but to do serious work on them we usually use another programming language to prepare and execute SQL commands automatically.

To talk to PostgreSQL from Python, the library *psycopg2*[4] can be used. This maintains two python variables representing a *connection* to a database (i.e. the state of your whole session of interaction), and a *cursor*, which contains your current position in, and results from, the database. To set this up in Python (e.g. typing at Spyder's IPython command line),

```
import psycopg2
con = psycopg2.connect(database='mydatabasename', user='root')
cur = con.cursor()
```

The most basic way to read from the database into a Python list, and print out the result, using *psycopg2* is like this (we will later use the Pandas library to replace this simple method),

```
sql = "SELECT * FROM Detection;"
cur.execute(sql)
mylist = cur.fetchall()
```

To execute a SQL insertion command (or anything else that changes the database), we both "execute" and "commit" it. This is called a "transaction" and is a safety feature. Suppose we were running a bank account database, and want to transfer money from account *A* to account *B*. If we gave two commands in sequence, to remove money from *A*, then to add it to *B*, and the power went down after command *A*, then this money would vanish from the system. With a transaction, we can execute several commands like this together in a "safe copy" of the database, then only put the complete results into the live database when they are all done, using the commit,

```
sql = "INSERT INTO Detection \
       (camera, licencePlate, confidence, timestamp) \
       VALUES (9, 'A06 NPR', 0.78, '2014-04-28:10:18:21');"
cur.execute(sql)
con.commit()
```

For an example of a Python script which uses this to load a CSV file, parse it, and insert the data into PostgreSQL, see *PySQL.py* in the Docker image. One of the main, tedious, problems here is usually in changing the formats of variables as we move between Python and SQL, for example you will see many commands that reformat dates and times. We will talk more about this kind of "data munging" in the next chapter.

3.3.5 Importing Vehicle Bluetooth Data

Derbyshire County Council maintains a network of Bluetooth sensors around its road network, which detect and log the unique identifiers of Bluetooth-enabled devices such as mobile phones and car radios

---

[4] Pronounced "psycho-P-G-two". "psyco" is the name of an internal piece of software inside some versions of Python; "PG" is for Postgres.

which drive past them. Bluetooth is a short-range radio communications protocol, used typically for streaming music from mobile phones to car radios and other devices. Part of the Bluetooth protocol consists of every device periodically publicly transmitting its unique identifier code, called a "Bluetooth MAC[5] address". For transport analysis purposes, detections of these unique IDs behave similarly to ANPR data. They can be used to track individual vehicles around the network by following the trail of the detections (but cannot identify the vehicles IDs against other data such as owner names, unless fused with additional linking data). Here we will process a log file from one sensor for one day.[6]

Tasks:

The locations of all sensors are given in the file,

~/data/dcc/web_bluetooth_sites.csv

Detections for a single sensor for a single day are given in,

~/data/dcc/bluetooth/vdFeb14_MAC000010100.csv

Each detection consists of a time, sensor site, and vehicle, and each site has a location and name.[7]

Design a database structure to model this data. Implement your database structure with PostgreSQL commands.

Write Python code to parse the DCC data and insert it into PostgreSQL.

Use SQL queries in PostgreSQL to query the database and report some interesting findings.

3.4 Further Reading

The original source for the relational model:

- Codd (1970) A relational model of data for large shared data banks. Commun ACM 13(6)

A good textbook covering database design and systems, and SQL programming is,

- Connolly, Begg (2014) Database systems: a practical approach to design, implementation, and management. 6th edn. Addison Wesley

There are many "teach yourself SQL in 24 h" type tutorials on-line and in programming books. These tend to focus on just teaching programming in SQL without thinking so much about the conceptual issues behind it.

If you want to learn more about philosophical ontology, an excellent place to start is,

- Roger Scruton (2004) Modern philosophy: an introduction and survey. Pimlico.

---

[5]"MAC" is from "Media Access Control".

[6]At the time of writing, some phone manufacturers design their phones to emit the same ID every day, while others have begun to randomize them once per day specifically to make long-term tracking of their users more difficult (but not impossible!).

[7]The Bluetooth detections supplied with *itsleeds* are based on real Derbyshire County Council data, but have been hashed in the public *itsleeds* release to prevent identification of individuals. The hashed versions look similar to real Bluetooth data, and will give a roughly accurate picture of real Derbyshire traffic, but may overestimate matching pairs of vehicles due to the hashing process. Students on the live ITS Leeds course may have access to unhashed data under data protection agreements to build more accurate models.

Data Preparation

In the practical examples so far we have read data from CSV files, placed it into SQL queries, and inserted it into a database. In general, this three step process is known as ETL for Extract, Transform, Load. Extract means getting data out of some non-database file. Transform means converting it to match our ontology and type system. Load means loading it into the database. ETL is often performed on massive scales, with many computers working on the various steps on multiple data sources simultaneously. For example, this happens when a transport client sends you a hard disc with a terabyte of traffic sensor data on it. In the "big data" movement, the transformation step might not be so important, as the philosophy here is to worry about ontology only at runtime, and store the data in whatever form you can manage when it arrives. As such, the Extraction step tends to be the most important and involved. Extraction is also known by various names such as "data munging", "data janitory", "data cleansing" and "data wrangling". It is a low-level but very important skill, which can easily take up half or more of the time of working data scientists. Because it is so important we dedicate this chapter to learning how to do it properly. It is often done quite badly, which can lead to very expensive problems later on, such as when your client realizes that your entire million dollar analysis is wrong because you accidentally swapped the days and months of American-formatted event data. The most famous data format disaster was the Millennium Bug, which has been estimated to have cost hundreds of billions of dollars to clear up, and arose from decisions to omit two text characters from each formatted date, resulting in the year 2000 being mistaken for the year 1900.

4.1 Obtaining Data

When dealing with large data sets we need to use large SI (metric, not binary) units as shown in Table 4.1.

Consumer hard discs in 2018 are terabytes; a big compute cluster is petabytes; the world's biggest company data centers are exabytes. All the data in the world was estimated to take around three zettabytes in 2017. The most efficient way to move terabytes or larger data around the world is to fill hard discs with data and put them on a truck or plane. We often move research data between countries by mailing parcels of hard discs. Amazon recently launched a truck-based data transport service, "AWS Snowmobile" which will physically collect data and upload it to their own servers. Smaller than terabyte data can be moved over the internet. Many governments now make public data freely available on their web pages such as *data.gov.uk*. Many countries including the UK also have

Table 4.1 Large SI prefixes

Name	Symbol	Value
kilo	k	10^3
mega	M	10^6
giga	G	10^9
tera	T	10^{12}
peta	P	10^{15}
exa	E	10^{18}
zetta	Z	10^{21}

"Freedom of Information" acts which allow you to force governments to give you most of their off-line data. Common ways to transfer data over the internet include:

- manually visit the web page and click links to download single files via your browser.
- *wget* command, performs the same *http*-based download as a manual click but programmatically.
- *ftp/sftp* are non-web protocol commands purely for file transfers.
- *rsync* is a tool primarily designed to synchronize data on two remote machines, by transmitting only changes between them to reduce bandwidth. If the server is running an *rsync* service, it can also be a nice way to download.
- For scraping data from really difficult human readable web sites (see below), Selenium is a tool which automates mouse and keyboard movements to "spoof" a human user.

4.2 Basic Text Processing

Many simple text files can be handled without any advanced tools. We have seen several built-in Python functions that perform useful elementary text processing, such as,

```
for line in open('filename.csv'): #reads one line of text at a time
  line=line.strip() #removes white space from ends of lines
  fields = line.split(',') #split line into strings separated by ','
  text = fields[0].upper()   #convert first string to upper case
  print(text)
```

The main rule of this type of processing is: *don't hold whole files in memory, work with one line at a time*. If you try to read the whole whole and then write it to the database you will often max out your computer's memory. Python and other languages' line-based read commands make it easy to step through one line at a time – use them.

- What are the limitations of this approach? Can you parse the Derbyshire data with it?

4.3 Formal Grammar: The Chomsky Hierarchy

The nearest thing we have to an academic theory of data munging comes from formal grammars, which are classified by the Chomsky hierarchy. Historically they were used in proving theorems about what different classes of computing device were capable of doing, with the Turing Machine being the most general computer, and other automata such as Finite State Machines and Pushdown Automata having distinctly weaker powers. You might have seen linguistics books try to write down the "rules" of grammar for languages like English, with forms such as,

4.3 Formal Grammar: The Chomsky Hierarchy

```
SENTENCE -> NOUN-PHRASE, VERB-PHRASE, NOUN-PHRASE
NOUN-PHRASE -> DETERMINER, ADJECTIVE, NOUN
NOUN-PHRASE -> NOUN
VERB-PHRASE -> ADVERB, VERB
DETERMINER -> the, a
NOUN -> car, truck, road, driver
VERB -> overtakes, parks, drives
```

The rules above form part of a formal grammar which defines a set of strings, called a language, that can be generated from them. Equivalently, the grammar also specifies the behavior of a machine which is said to "accept" the language if it can process and recognize all strings from it. Different types of machine can implement acceptance of different types of grammars.

4.3.1 Regular Languages (Type 3)

The simplest (and least powerful) type of grammar in the Chomsky hierarchy are "Regular Languages". These are generated, and accepted, by Finite State Machines, which can be viewed as a network of nodes representing states, which emit (or accept) characters as we move between them. The FSM shown in Fig. 4.1 can output (or accept) strings such as: "a", "ab", "abbb", "abbbbbb". (It could also get stuck in the bottom right state, but strings are only output if we reach the "final state" and decide to stop.)

- Can you design an FSM to accept UK licence plate numbers?

FSMs are often used in robot control applications such as self-driving cars, but are usually impractical to work with directly to parse text files. Instead, languages such as Python and Bash provide tools to express equivalent recognizers called Regular Expressions, often abbreviated as "regex" or "regexp". Internally, these may be compiled into FSMs, but externally they show a simpler syntax to the programmer. For example the language above would be described by *ab** which means "one a followed by any number of b's (including zero)". The star is sometimes called the "Kleene star" after its inventor.

ab+ means "*a* followed by one or more *b*'s"

a[bc] means "*a* followed by either *b* or *c*" (sometimes written *a[b/c]*)

a([bc])* generates/accepts strings such as: *abcbcbc, abbbbbcc, abbb, accc, a*.

Fig. 4.1 A finite state machine (FSM)

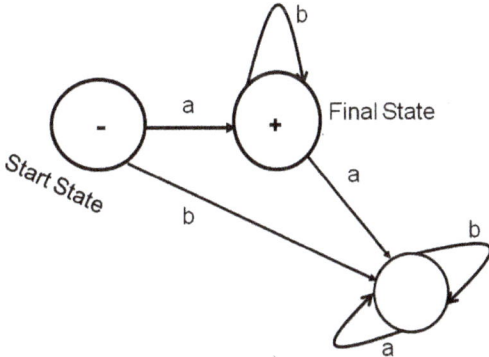

Designers of regex tools have agreed various other standard notations including "." to match any character, and \d or *[0-9]* to match any digit, \s to match any whitespace character and \S for any non-whitespace character.

For example, suppose we have a text file which has lines containing pairs of property names and values such as:

```
name: car1 , speed: 65.5 , driver: alan
name: car2 , speed: 62.5 , driver: stephen
name: car2 , speed: 62.5 , registration_year: 2002, driver: noam
```

We can model and recognize such lines with a regex like *(([a-z]:\s*\S+\s*,\s*)+)*. Regexes can look scary but are a very concise and practical way to describe regular languages – much easier than implementing FSMs directly.

In Python, regexes are implemented by the *re* library. They also exist (with slightly different syntax) in Bash commands, in most other languages (especially Perl, which makes heavy use of them), and the command line tools *sed* and *awk*.

4.3.2 Context-Free Languages (Type 2)

Sometimes, FSMs and regex's are not powerful enough to describe your input data files, and you need to move to the next most complex language model. A common case where this occurs is when your data arrives in some hierarchical structure, where you need to keep track of symbols like brackets and braces and ensure that they match up. For example, an FSM cannot recognize a language of strings like *(((hello))), (hello (world)), (car1(driver(alan), speed(60)), car2(driver (person(name(stephen), age(27)))*. This type of representation is common in structured and object-oriented (non-relational) data languages like XML and HTML.

In computation theory, the type of machine that can accept these is called Pushdown Automaton. This is like a FSM with an added "memory store", which can be used to keep track of the brackets seen in the past. As with FSMs, we don't usually implement pushdown automata directly or care about how they work (one way to imagine or implement them is as a kind of recursive FSM, where each state in the FSM can expand into another complete FSM), but use an equivalent structure to describe them more easily, called a *Context Free Grammar (CFG)*. The simple English model in the previous section is a (partial) example of a CFG. Its defining characteristic is that the left hand side of each rule contains exactly one symbol, while the right hand side may contain many. By specifying a list of such CFG rules, a programmer can pass them to a software parsing tool, which will compile them into a pushdown automaton-like compiler, and use them to recognize the language automatically. This can be extremely powerful, for example just by inputting the English grammar above, we can understand the structure of a large number of English sentences, which could be used, for example, to implement a natural language or speech recognition interface to a mobile phone traffic app. Another common use is for "scraping" data from HTML web pages, using grammars such as,

```
WEB_PAGE = HEAD, BODY
BODY = DATA_TABLE*
DATA_TABLE = COLUMN_HEADS, ROWS
COLUMN_HEADS = (fieldName,)*
ROWS=ROW*
ROW = (fieldValue)*
```

4.3 Formal Grammar: The Chomsky Hierarchy

You will typically do this after retrieving the text of of web page programmatically with a command such as,

```
$ wget http://www.mysite.com/index.html
```

Please be aware that scraping other people's web data in this way might not always be legal or ethical! In particular, some websites may try to detect you doing this, and take countermeasures, such as banning you from their servers or (more deviously) silently switching to feeding you with false information. For example, if I am providing a commercial traffic data application, I want human users to see my data, but I don't want you to steal it all and use it to provide a competing service. Then you might take counter-counter-measures to disguise your data theft attempts, and so on. There are very nasty – though often highly paid, and enjoyable to those of a certain mindset – cat-and-mouse games which can be played here. In particular if you are providing a public web application with an database back-end yourself then you will need to take extra security measures against "injection attacks" in which an attacker uses punctuation marks in their input strings to inject runnable code into SQL queries, potentially giving them full access to your database (Fig. 4.2).

4.3.3 Beyond CFGs (Types 1 and 0)

Is is unusual to need more power than CFGs for normal data extraction. The higher level grammars are usually only used when considering complex languages such as computer programming languages, music, and natural language. However – modern Data Science is increasingly interested in extracting information from natural language, such as social media feeds, so they may become more popular. To give a brief overview: Context-Sensitive Grammars (CSG) (Type 1) can be written with rules which look similar to CFGs, but allow pattern matching on the left side as well as substitution. i.e. whether a rules is allowable depends on the context around it. For example, these rules make *NOUN-PHASE* expand differently for definite ("the") and indefinite ("a") articles,

```
the NOUN-PHRASE -> the ADJECTIVE, NOUN
a NOUN-PHRASE -> a NOUN
```

Most generally, "Recursively Enumerable" languages (Type 0) are those which are generated and accepted by arbitrary computer programs (Turing machines). Unlike with the other grammars, computation theory comes into play at this point and can show that it is impossible make these work perfectly, because it is always possible to create cases which cause the computer to go into an infinite loop rather than give any answer.

Fig. 4.2 From *www.xkcd.com*

Fig. 4.3 Floating point storage

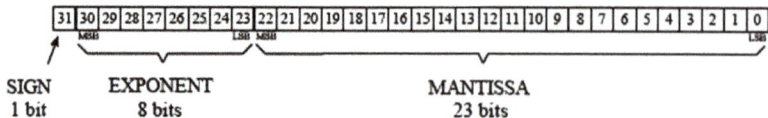

In practice, most "regular expression" and CFG libraries actually implement aspects of machines from higher Chomsky levels than their name suggests. It is common to find CFG and CSG features bolted on to them.

4.4 Special Types

4.4.1 Strings and Numbers

Let's talk about text. While high-level ontology deals with representations of relationships between entities, all of their properties will ultimately bottom out into a few basic data types such as strings and numbers. It is important to get these right because they can massively affect the speed at which your system runs. Strings are represented as sequences of characters. In most systems, each character is represented by one byte (8 bits) using the ASCII code, such as the letter 'A' by the number 65 (binary 01000001). Most languages have functions to convert between char, byte, integer and binary representations. If you are dealing with data including non-Roman characters, it may use the newer Unicode standard, which uses sequences of bytes to represent thousands of international alphabetic, mathematical and other characters. Strings are easier to work with if you know they are of fixed lengths and have information in fixed positions. For example in Python you can read the 9999th character of a string almost instantly using *mystring[9999]*. In contrast, if you need to search through strings (e.g. using regexes) it will take much longer.

Numbers are typically represented on PCs and servers in two ways, integers and floating-points. An integer (nowadays) is usually 32 or 64 bits long, with all but one bits used for the number and one bit sometimes for its plus/negative sign. Floating point numbers are more complex, representing numbers like $3.44645635 * 10^{34}$. Rather than store all the significant figures, they store the most significant ones (mantissa) plus the exponent and sign as shown in Fig. 4.3.

(Floating point computation is quite complex for processors and is often missing on embedded computers as found in sensor networks and ANPR cameras.)

Often your data will arrive in a text format, where integers and strings have been represented by ASCII characters, such as "45.567" for 45.567. To a computer, these do not behave like numbers at all, their binary values bear very little relation to their arithmetic properties. So you will need to convert them to actual numbers with casts, for example in Python,

```
val = int(field)    #cast a string number to an integer
val = float(field)  #cast a string number to a floating point value
```

A common error is to forget this, or do it wrong, leading to the ASCII integer values of the characters, such as 49 for "1", ending up in arithmetic.

4.4.2 Dates and Times

In the author's experience, formatting dates and times can occupy 50% of data scientists' time in real life, especially when other people have done it wrong, so let's get this right from the start!

4.4 Special Types

Dates and times have a complex structure and history. Many time standards arose due to transport needs. Before the railways, each UK city would maintain its own standard defined by reference to physical noon, i.e. the instant the sun reached it highest peak in that city. Greenwich Mean Time (GMT), was defined such that 12:00 was physical noon at the Greenwich Meridian line in London each day, while Oxford time ran 5 min and 2 s after GMT, when the sun is highest was Oxford. The GMT standard was adopted across the UK (with the single exception of Christ Church, Oxford, which retains Oxford time) to enable railway timetables to function in a single time zone.

Now that the world is globalized, we face similar problems converting between larger time zones around the planet. Coordinated Universal Time (UTC) is the modern global standard which is *almost* synonymous with GMT in practice, though defined precisely by reference to atomic clocks rather than to physical noon at Greenwich. World time zones are usually related to UTC by shifts of integer hours, but not always. For example Iran, Afghanistan and North Korea have fractional hour shifts from UTC.

Dates and times are ultimately intended to refer to (though are no longer defined by) physical motions of astronomical bodies, which are not nice and regular. The basic hours and minutes structure of time used today comes from ancient Mesopotamia, from at least 1,500BCE, where it was common to do arithmetic in base 60. (This is actually much nicer than base 10 because it factorizes in many more useful ways, including 5×12.) To stay synchronized with the astronomy, dates and times move in strange ways during leap years, leap seconds, and other minor adjustments. If one of your systems corrects for these and the other does not, then bad things will happen. For example, estimating the speed of cars detected by two ANPR cameras where one has not included a leap second could make the difference between legal and illegal speeds, and prosecution. If you ever need to work with centuries old data (e.g. populations and climate models), remember that calendar standards have changed a few times over history, such as when the English removed 11 days from 1752, so a good date library must include these conversions. When dealing with GPS localization, relativistic effects can become non-negligible – time itself passes more slowly on a fast moving satellite than on earth, with potential effects around the same size as RTK precisions! In other applications, the difference between week days and calendar days is important. For example when scheduling work, we want to count in "business days", of which there are about 250 in a year rather than 365.

Various people, such as French Revolutionaries, have tried to simplify these structures with decimal and other representations, but all have failed for everyday use. The SI unit for time is, however, just seconds, and a common way to represent dates and times together in computer data is the number of seconds since the "epoch", which occurred at the start of Jan 1, 1970, defining the "birth" of the computer age.

Most modern systems will isolate the programmer from the internal representation of dates and time by providing a pre-built *Date*, *Time* and *DateTime* data types. These might, for example, represent time since the epoch internally but allow to you talk to them via human readable formats such as "01-03-2014 09:35:33.565 CET".

Time is different in different places in the world, including where your data was collected, where the sensor was manufactured, where your computer was made, where the server your data is stored on, and where you and your cloud server are both working (in the cloud, you might not even know what country your server is in). Many regions switch between two time standards for summer and winter such as GMT/BST in the UK. They do not do this all on the same day. Ways to handle time zones include:

- store all times as the standard UTC,
- store all times in the standard where the data was collected,
- store each time with an extra field specifying the time zone.

A particular danger for transport systems is designing them based on local time for use in a single location, then trying to export them for use in other countries for profit when they work. Again, internally, many tools will take your time zones and convert everything to UTC or epoch time, but you need to think about conventions when you interact with them. Use your programming language's libraries – don't try to do conversions by yourself!

The most dangerous time issue is USA date formats, which are traditionally written MM-DD-YYYY instead of DD-MM-YYYY. This can very seriously ruin your system and your career because it may go un-noticed and lead to importing terabytes worth of meaningless rubbish! *The only ways you should ever write dates are "2014-01-15" or "2014-Jan-15". If you only learn one thing about data munging, this should be it!*

4.4.3 National Marine Electronics Association (NMEA) Format

Satellite positioning (including GPS) data is of particular interest in transport. We will talk more about this data in the next chapter, and just take a brief look at its format here.

Most physical sensor devices come with serial port which streams lines of text to a computer, as a stream of bits chunked into ASCII characters (bytes). If you insert a probe into their serial port cables you can actually see the 0s and 1s over time in the voltage. The format of this data from most GPS-type devices is called NMEA and looks a lot like a CSV file, with lines like,

```
$GPGGA,123519,4807.038,N,01131.000,E,1,08,0.9,545.4,M,46.9,M,,*47
```

You will often receive log files with lots of these lines recorded over time. The first symbol is the type of data. GNSS is very complex and produces lots of information about individual satellites and accuracies, but the lines which begin with "GPGGA" give the best summaries of the actual location as you would see reported in a car satnav. The second field is a time stamp in HHMMSS string format, using UTC time. The third and fifth fields are the latitude and longitude, in degrees, as strings. You will typically filter NMEA log files to get just the GPGGA lines, then pull out these three fields.

4.5 Common Formats

As well as text and CSV files, you will sometimes get structured web data files such as XML, JSON and HTML. "Web 3.0" is/was supposed to be a world where semantic structured data, rather than just human readable web-pages, is transferred around the internet. It has never been very well defined though, and most efforts run into the ontology problems we discussed previously which prevent them from generally talking to each other. However there are now several standard ontologies for specialist domains, mostly written in XML, which can be useful. XML uses the object-oriented (not relational) model to serialize data from hierarchical objects into an HTML-like syntax. For example,

```
<guestbook><guest><fname>Terje</fname><lname>Beck</lname></guest>
<guest><fname>Jan</fname><lname>Refsnes</lname></guest><guest>
<fname>Torleif</fname><lname>Rasmussen</lname></guest><guest>
<fname>anton</fname><lname>chek</lname></guest><guest><fname>
stale</fname><lname>refsnes</lname></guest><guest><fname>hari
</fname><lname>prawin</lname></guest><guest><fname>Hege
</fname><lname>Refsnes</lname></guest></guestbook>
```

JSON does the same thing but without looking like HTML, for example,

4.5 Common Formats

```
{"menu": { "id": "file", "value": "File", "popup":
{ "menuitem": [ {"value": "New", "onclick": "CreateNewDoc()"},
{"value": "Open", "onclick": "OpenDoc()"},
{"value": "Close", "onclick": "CloseDoc()"} ] } }}
```

There are standard parsing tools available for these formats which mean you don't have to write your own CFGs – for Python see the *Beautiful Soup* library.

If you are working with object-oriented code, these are nice formats to read and write data, and most languages have libraries to do this automatically. However – as discussed in the Database Design chapter – mapping between objects and relational data can get quite messy. A common approach is to restrict your relational model so that each relation describes one class, and each row in its table is one object. If you do this then there are libraries which will automatically translate between SQL and objects in many languages. (Though you lose the power of the relational model to chop and join relations of course.)

Some data arrives in unstructured, human communication forms, such as English sentences in bulk social media conversation data, or audio recordings from call centers. Modern speech recognition for limited domains and speakers can approach human transcription levels, turning audio into natural language text. But turning natural language into structured data is hard. Parsing methods, though theoretically pretty, tend to break in practice. So simpler statistical "frame-based" methods tend to be used. For example, we might (as human programmers) define particular stereotyped events such as "booking a train ticket" or "reporting a traffic jam". These can then be spotted by looking for certain keywords, and numbers or values pulled out from simple rules around the keywords. "Real" parsing based language processing remains an active research area and it is frustrating that it is not competitive with these more "munging" style methods – however IBM's Watson claims to be making progress by mixing it with the statistical frame based approach (e.g. search for the report *Vehicle telematics analytics using Watson Internet of Things Platform Cloud Analytics*).

4.6 Cleaning

Once you have imported data, it is important to inspect it and fix any remaining glitches. For inspection, it is useful to run various "sanity checks" to ensure the data has the properties you think it should have. For example, compute the mean and variance of properties, or plot their distributions. Compute or plot correlation structures between properties. Plot values over times and space. Work with your client to understand the data and define checks. This can be a good time to explore and play with the data to get ideas for what to do with it too.

As well as not seeing structure where there should be structure, another warning sign of a bad import is seeing structure where there should not be structure. If you expect a variable to be "random" in some sense and it shows predictability then something might have gone wrong. A common source of this is importing data after incorrect type conversions, which may act on the binary data to introduce structure. There are various "randomness detector" tools which try to give you clues about when data is not random, by applying banks of standard statistics to it. (Note it is not possible to prove completely that data is "random", only when it is not, because there could always be some more complex statistic to try that would show structure.)

A common problem is missing data. This may appear as a "null" or "NaN" ("Not a Number") value, and play havoc with arithmetic operations such as computing means. A single "NaN" will explode and make all calculations based on it become NaNs too. Handling missing data is notoriously tricky. Approaches include:

- Set all the missing values to 0 (bad because it will skew statistics which include them)
- Set all the missing values to the mean of the good values (better, but will skew variance and other statistics)
- Estimate the values of all the statistics you care about, and sample the missing variables from this distribution. (This is a simple form of the EM algorithm. It is often a useful method for transport.)
- Leave them as NaNs and ensure all your algorithms have specific alternative methods for handling them. (This requires the most work.)

The last approach is often required for financial systems, such as when using transport data to model customer behavior and predict sales and stock prices from it. This is because financial data is used by extremely competitive machine learning researchers who must be assumed to be able to find and exploit any possible patterns in the data. Any attempt to fill in missing data using an algorithm will introduce some artificial structure, which the learning algorithms may then pick up on and exploit, to give good-looking but fake performance in historical trading simulations. It is also sometimes the case that a real-world system will have to handle missing values anyway – for example when a stock exchange suspends a stock it is really not possible to trade it any more.

4.7 B+ Tree Implementation

Most of Database Systems design can be left to the Computer Scientists who write database programs like Postgres. Different programs use different structures to represent the data. However almost all make use of a particular structure, the "B+ tree" which it is useful to know about to speed up your SQL queries. Recall that if we store N data in flat lines of text and want to search for an item, it takes time $O(N)$ to find it. (The "Big O" means "order of"). However, if ever datum has a unique, ordered, key such as a licence plate or citizen ID, then we can form various types of tree structures which represent the ordering of the keys. Using trees, it is possible to locate an item in $O(log(n))$ time. The B+ tree is a particular type of tree as shown in Fig. 4.4.

- Try searching for item "65" – how many steps does it take?

To tell a database to make use of such structure, we create *indices* on our tables. This can be done in the SQL *CREATE* commands, or afterwards once the data is loaded. We saw examples of indices as *PRIMARY KEY* commands during *CREATE*s in the last session. It is possible to put indices on as many fields as you like, not just the primary key. However each index may require significant space

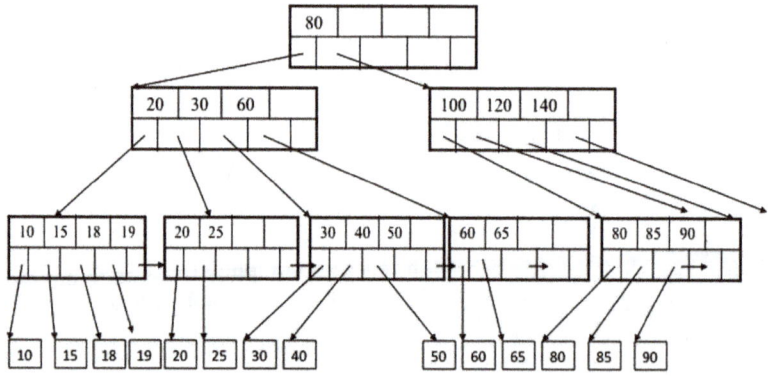

Fig. 4.4 B+ tree

on your hard disc, and take a time (e.g. weeks) to create. If you are hired to speed up a database, this can give very quick, easy and highly paid wins!

Indices are a form of redundancy built into databases to trade storage space for speed. There are other forms of redundancy too. While relational theory recommends not to store duplicate information, "data cubes" (aka "OLAP cubes", for On-line Analytic Processing) are often used, which are extra tables set up to cache data from multiple normalized tables in one place.

4.8 Exercises

4.8.1 Reading the Database with Pandas

So far we have used Python to insert data into the database but not yet to read it. There are several different ways to read, with new wrappers still appearing sometimes, including:

- The most basic command is *cur.fetchall()*, which will return the result of a query as a basic Python list
- The Pandas library provides a higher level interface to read data directly into data frames, as in the following example,

```
import pandas as pd
con = psycopg2.connect(database='mydatabasename', user='root')
sql = 'SELECT * FROM Detection;'
dataFrame = pd.read_sql_query(sql,con)
```

Try retrieving data from the previous chapter's exercises using Pandas.[1]

4.8.2 *printf* Notation

A useful Python syntax known as "printf" notation[2] allows you to insert variables into strings like this, and is especially useful for working with SQL strings,

```
my_str="michael knight"
my_int=32
my_float=1.92
s = "name: %s, age: %d years, height: %f"%(my_str, my_int, my_float)
```

Try some of the exercises from the previous chapter using this syntax.

---

[1] Pandas has many powerful features to perform SQL-like operations from inside Python, and to assist with data munging. For direct code translations between SQL and pandas, see *http://pandas.pydata.org/pandas-docs/stable/comparison_with_sql.html*. Or you can use converters like *pandasql* which let you run actual SQL syntax on their data, without using a database. Also refer to the table in Chap. 2 for some useful Pandas commands for Transport applications.

[2] Pronounced "print F", from the *printf* command in C-like languages.

4.8.3 DateTimes

Here are some basic Python commands for processing dates and times with Python's *datetime* library. (A full tutorial is at *www.tutorialscollection.com/python-datetime-how-to-work-with-date-and-time-in-python/* (Note that DateTimes are called "DateTime" in Python and "timestamp" in PostgreSQL.)

```
import datetime
#make datetime from string
dt = datetime.datetime.strptime('2017-02-11_13:00:35.067' , \
                                "%Y-%m-%d_%H:%M:%S.%f" )
#convert datetime object to human readable string
dt.strftime("%Y-%m-%d_%H:%M:%S")
#create a TimeDelta object for difference between two datetimes
delta = dt-datetime.datetime.now()
#or from scratch:
delta = datetime.timedelta(milliseconds=500)
#print delta as a human readable string
str(delta)
#convert delta to float number of seconds
delta.total_seconds()
```

Here is a common Python-SQL idiom to recover the latest value of something up to a time,

```
sql="SELECT * FROM Detection WHERE timestamp<'%s' \
    ORDER BY timestamp DESC LIMIT 1;"%dt
```

4.8.4 Time Alignment and Differencing

Sometimes you may need to do this in bulk – aligning all values of one time series to another. Iterating many commands like the above SQL is possible but slow. It is possible but quite complex to do this as a single SQL query, but much easier using Pandas' re-sampling and join methods when the timestamp column is indexed (here we create an index in Pandas rather than in the database),

```
df_A = df_A.set_index(['timestamp'])
df_B = df.B_index(['timestamp'])
df_B = df_B.resample(df_A)
df_C = df_A.join(df_B)
```

For transport data, it is very common to compute speeds (and accelerations) from time series of positions, using Pandas like this (but note that it introduces a NaN as the first speed),

```
df['dt']=(df['timestamp']-df['timestamp'].shift(1))/np.timedelta64(1,'s')
df['speed'] = (df['pos']-df['pos'].shift(1))/df['dt']
```

Create some tables in your database, load DataFrames from them, and use them to test re-sampling and speed computations.

4.8.5 Parsing

Using Python's *re* regex library, the following matches some example data line strings,

```
import re
print(re.match("(\d+) , (\d+)", "123 , 456").groups())
```

This gives a list of all the matches from the brackets, in this case: "123" and "456".

Try to match sentences like "Hello Michael" to extract the person's name. Make it match names that have a first name and a second name like "Michael Knight". Then add optional titles and middle initials such as "Mr. Michael A. Long".

Try to match and extract lines from the Derbyshire Bluetooth data from the previous chapter, this time using *re.match*. There is a full description of its syntax here: *https://docs.python.org/2/howto/regex.html*. Is this faster or more pleasant than using the previous raw text operations?

Regexs are also very useful using the terminal command-line (not IPython) tool *sed* if you want to quickly search and replace a file,

```
$ #replace 'long' with 'knight' everywhere. "|" is called a "pipe".
$ cat input.txt | sed 's:long:knight:' > output.txt
```

Try parsing simple greetings like "Hello Michael" using a PyParsing CFG,

```
from pyparsing import Word, alphas
greet = Word( alphas ) + "," + Word( alphas ) + "!"
greeting = greet.parseString( "Hello, World!" )
print(greeting)
```

Now try parsing a GPS string using CFG parsing from PyParsing,

```
from pyparsing import *
survey = "'GPS,PN1,LA52.125133215643,LN21.031048525561,EL116.898812"'
number = Word(nums+'.').setParseAction(lambda t: float(t[0]))
separator = Suppress(',')
latitude = Suppress('LA') + number
longitude = Suppress('LN') + number
elevation = Suppress('EL') + number
line = (Suppress('GPS,PN1,')+latitude+separator+longitude+separator+elevation)
print(line.parseString(survey))
```

Try to build a tiny natural language parser that looks for tweets like "The [number or time] [bus or train] is [late or on time or early or busy or full]".

4.8.6 Vehicle Bluetooth Munging

As in the Chap. 3 exercise, create database tables and import the Derbyshire Bluetooth data, this time using Pandas to import, and using Python DateTime and SQL timestamp types to store DateTimes correctly. Write a loop to import data from all the Bluetooth sensors rather than just one.

4.9 Further Reading

The classic approach to handling missing data. As with Codd, few people read it all but everyone should at least glance at it once in their career,

- Dempster AP, Laird NM, Rubin DB (1977) Maximum likelihood from incomplete data via the EM algorithm. J R Stat Soc Ser B 39 (1):1–38
- Or try the Wikipedia page, *Expectation–maximization algorithm*. The algorithm is actually very simple, you just fill in the missing values with your best guess of them, then fit your model again, and iterate. That's all!

If you would like to learn more about how B+ trees implement indices,

- https://www.youtube.com/watch?v=h6Mw7_S4ai0 (animation of the algorithms in action)

More details on regular expressions can be found in,

- Friedl JEF (2006) Mastering regular expressions. O'Reilly, US

For a full mathematical treatment of languages, automata, and Chomsky hierarchy, including Turing's famous computation results,

- Lewis H, Papadimitriou CH (1997) Elements of the theory of computation. Prentice Hall, Upper Saddle River

Spatial Data   5

Transport data is generally about motion through time and space. The previous chapter considered the complexities of representing time – here we will think about space.

5.1 Geodesy

Questions about transportation usually concern distances between meters and thousands of kilometers. At the lower end of this scale, we can use a simple Cartesian grid of meters without any problems, for example to model the layout of a city. At the higher end of this scale, the curvature of the Earth's surface becomes an important factor in representing the data.

A *datum* is a model of the Earth's shape.[1] A naïve model of the Earth assumes it is a perfect sphere, whose positions are uniquely (though redundantly at the two poles) described by two coordinates, latitude (north) and longitude (east). However the real Earth is not a sphere. Most basically, its global shape is more like an ellipsoid. Then rather than being a perfect ellipsoid, it has an uneven surface, with bulges at different places. There are many possible choices of geometric object that can be used to approximate this shape, which define their own latitude-longitude coordinate systems. Some are more accurate than others for particular regions of the Earth's surface. If you receive data notated in different datums then you will need to convert them into one standard format. The most common global datum is called WGS84 (World Geodetic System 1984). The ISO6709 standard defines that we should always write ellipsoidal coordinates as (latitude, longitude), sometimes called "latlon", not the other way round.[2]

Projection. It is often more convenient to work with a flattened 2D projection of a datum rather than its spherical coordinates. Usually this happens where we are interested in some region such as one city, but have received data from a national-level collection using global coordinates. Then we want to pick some point, such as the center of the city, as an origin, and project the coordinates into Cartesian x and y meters. The standard convention for such projected maps is to take $x =$ Easting and

---

[1] This is a different meaning from the occasional use of "datum" as the singular of the Latin plural "data".

[2] This is somewhat unfortunate because when dealing with (x, y) map coordinates we often want to have (horizontal, vertical) ordering. Some tools do violate the ISO standard and use lonlat for this reason. It occurs in *pyproj* and also if you want to plot spherical coordinates directly as *plot(lons, lats)*. Be careful as this is a common source of bugs!

Fig. 5.1 British National Grid

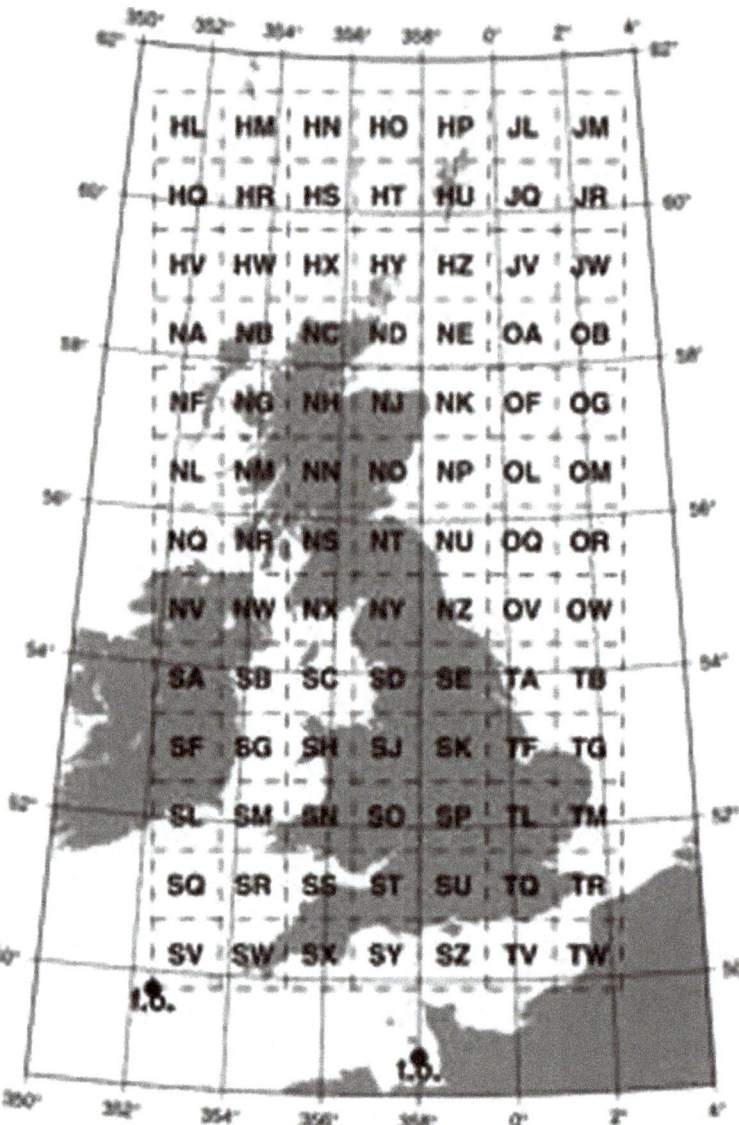

y = Northing, in the order (x, y), in meters from some origin.[3] It is not possible to make a perfect flat version of an ellipsoid surface, so any projection must make some compromise. Usually a good projection to use is one of the Universal Transverse Mercator(UTM) family. UTM splits the Earth's surface into state-sized regions, and defines separate projection for each one, to minimize the distortion there. UTM is used for example in the UK Ordnance Survey's National Grid[4] (also known as "British National Grid", BNG) coordinate system, which then divides the country into a hierarchical grid using letters and numbers, as shown in Fig. 5.1.

---

[3] When working with data represented in a vehicle's own frame of reference, known as "egocentric" coordinates, an emerging standard is x = "straight ahead", y = "left", ordered as (x, y). This can be confusing at first as you will often want to draw plots with x pointing upwards and y to the left, for example, when working with autonomous vehicle simulations.

[4] Not to be confused with the unrelated UK electricity distributor, National Grid Plc.

5.1 Geodesy

If you are working with inter-country and inter-continent data such as flight or shipping paths then you probably want to stay in ellipsoidal coordinates and not project at all. In such cases you may also need to consider height coordinates, and remember that one degree projects to different distances at different altitudes. An aeroplane covers a longer physical distance than the length of its map projection because it must climb and descend as well as move around the 2D map – as a passenger you may have seen displays showing distinct air speed and ground speed. Similar issues arise when working with ground vehicles when the inclinations of hills create roads which are physically longer than their map projections suggest.

If you try to merge data from two sources and notice that the roads don't quite line up, this is probably due to a datum/projection conversion issue. For UK data, this tends to produce characteristic shifts of a few meters.

5.2 Global Navigation Satellite System (GNSS)

Global Navigation Satellite Systems (GNSS), include the best known Global Positioning System (GPS, American) and also the alternative GLONASS (Russian), Galileo (EU), and BeiDou(China). GNSS can also refer to fusing data from several of these systems. Details of the systems change over time as new satellites are launched and upgraded so the following is just a sketch of one system. GNSS are important military assets which each country maintains in case the others are disabled. It is likely that more satellites will continue to be launched to build in redundancy to all systems as states also build anti-satellite weapons. The advantage of this for transport researchers is that in peacetime, we get access to four systems, each with lots of satellites, to fuse together and improve our accuracies.

The original GPS system consisted of 24 orbiting satellites about 20,000 km altitude, as shown in Fig. 5.2. From any point and time on Earth, one can typically see between 6 and 12 satellites in the *unobstructed* sky. Urban and other obstacle-filled environments are unlikely to see as many, which can be a big problem. The satellites transmit microwave (1.2–1.5 GHz) signals, containing their identity and correction information, at the speed of light, timed by atomic clocks. Receivers compare the time delays of signal arrivals, and triangulate their position, using knowledge of the satellites' positions and conditions. As the changing atmosphere affects the signals, a network of base stations at known

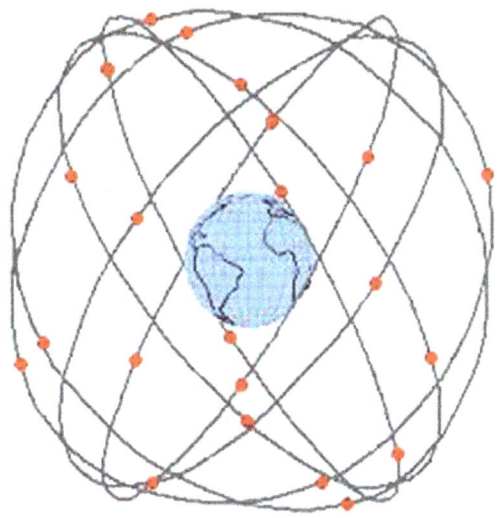

Fig. 5.2 GPS satellite constellation

Fig. 5.3 *RTK/DGPS base station*

locations monitors the errors, and computes and relays correction information back to the satellites, which include it in their transmissions. Localization accuracy on the Earth's surface from this basic system is around 10 m. The satellites move around slowly in the sky, for example taking 30 min to appear and vanish over the horizons.

Differential GPS (DGPS) systems give higher accuracy, of the order of 100 mm, by installing their own local base station in addition to the global network, and comparing the computed location with that of a moving vehicle. (You can't make a "poor-man's DGPS" from two standard GPS receivers however, because they must use exactly the same set of satellites, which requires extra logic for them to communicate and negotiate with each other.) Some wide regions have public DGPS services which transmit over radio such as WAAS in the USA and EGNOS in the EU.

Real-Time Kinematic (RTK) GPS (Fig. 5.3) uses the DGPS concept together with additional high-resolution carrier wave phases (Fig. 5.4) between signals (rather than delays in the carried signals) to obtain accuracies up to 20 mm on a good day. On a bad day, cheaper RTK sensors don't work at all. Some organizations operate local networks of RTK+DGPS base stations, such as *rtkfarming.co.uk* which covers the East of England using farmer volunteers' base stations.

For limited smaller environments, such as a single street, even higher precision can be obtained using Real Time Localization Systems (RTLS). These work on the same principle as GNSS but using locally installed radio beacons rather than distant satellites. Generally, GNSS is not accurate enough by itself to localize vehicles for autonomous driving, especially in urban areas, and is fused and filtered with many other sensors, including its own historical readings, for this purpose. Similarly for car satnavs, heavy use is made of smoothed sequences of observations and knowledge of its speed and heading. It is easier to predict and smooth a high-speed car's positions on a motorway than its finer movements around intersections. This is why expensive RTK is common for agricultural tractors but is less common for cars. A vehicle's heading can be estimated either from its direction of motion or from a pair of RTK sensors placed at its opposite ends.

5.2 Global Navigation Satellite System (GNSS)

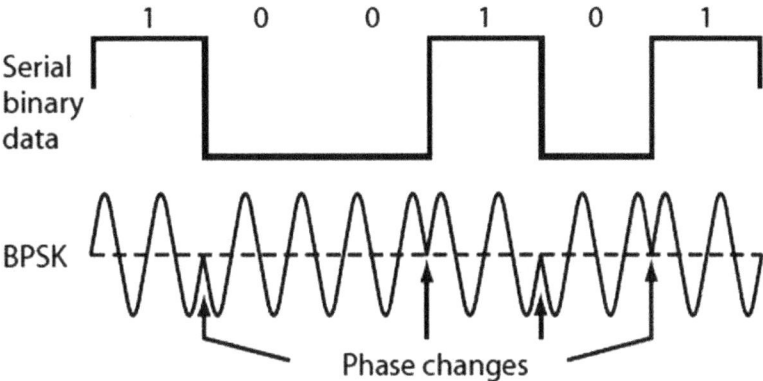

Fig. 5.4 *Carried (top) and carrier (bottom) signals*

Usually Data Science will use only the final estimates locations from NMEA data sentences (detailed in the previous chapter). Occasionally, data scientists may work with data from the individual satellites which also appear in the NMEA data, for example when researching new algorithms for fusing data from different systems or with other localization data sources.

5.3 Geographic Information Systems (GIS)

5.3.1 Role of GIS System

A Geographic Information System (GIS) is any system that is specifically designed to work with spatial data. It is possible to work with spatial data without a GIS, for example we could just store the latlon GPS co-ordinates of entities as pairs of *float* properties in Postgres. However if we do this, we will very quickly find ourselves needing to write lots of functions to do small and large jobs with this data, which tend to be the same in every project. GIS systems are implementations of these standard tasks, which may be packaged either as programming language libraries and/or as graphical user interfaces. Standard tasks include: converting datums and projections, searching quickly for entities in particular regions, searching quickly for entities with spatial relationships to other entities, mixing raster image data with vector data, handling lots of data at large and small scales (such as cities becoming point-like then vanishing as we zoom out from a map), specialist geographical data visualizations, and converting between standard spatial data file formats.

- You have probably used OpenStreetMap, Google Maps and similar web-based GIS tools. How much of the above work can they do?

Often if workers say they "do GIS" or "use the GIS" they mean that they have been trained to use some graphical interface software package which collects many of these tasks together and enables interactive use. These packages usually combine aspects of a database (e.g. Postgres), a CAD program (e.g. FreeCAD, SolidEdge), and drawing (e.g. LibreOffice Draw, Corel Draw) and painting programs (e.g. GIMP, Photoshop). The best open source GIS package is QGIS, available from *www.qgis.com*.

In Data Science we have a slightly different meaning: we will more typically be working with the same tools but packaged as programming language library functions instead of graphical interfaces. Sometimes the difference between the two is blurry, for example QGIS has the ability to link to Python programs to script its functionality. Usually the graphical interfaces just wrap the same libraries that we use directly.

Some questions that Data Scientists might ask of GIS systems:

- what site near the Derbyshire's A61 road has low road usage and high accessibility to commuters (to locate a new business park)?
- what origin-destination routes around the M25 motorway could be replaced by direct new roads or public transport links?
- what type of entity owns the most land across the EU that is suitable and profitable for extreme terrain agricultural robots?
- which stores have the highest traffic in their car parks (so we can profit by buying their shares before their sales figures come out)?

5.3.2 Spatial Ontology

Spatial objects have specific spatial properties. The philosopher Kant argued that certain fundamental properties are basic, or innate in our perception of the world, and that these include space and time. This provides an argument for treating them specially in our representations of the world. Like time, (perceptual) space is a continuous property. Unlike time, it is three dimensional. As we usually perceive one state of the world "at a time" there is another difference, that entities in space are perceived as *extended* over space but not (directly) extended over time.

With geographic data we typically work in only two of the three dimensions – on the ground. Two dimensional space supports three basic types of spatial entity:

- points – having a location,
- lines – comprising two or more (potentially infinite) locations in an ordered sequence,
- polygons – areas defined by three or more vertex locations in an ordered sequence.

Think carefully about these definitions. Points, lines and polygons are not as simple as they may first look. For example, both ontologically and computationally, should we think of a straight line as made of an infinite number of points, or as just the two points that define it? In the latter case, the infinite points connecting the two defining points have a different status than the defining points. Similarly for polygons, the status of the area being described is different from that of the defining vertices. In a real database system we cannot represent infinite sets directly and must work with the finite defining sets. So the entity being represented – the area – is not represented directly. So some serious computation may be required if we want to ask geometric questions about these entities such as:

- whether two polygons intersect (an "overlaps" relation),
- whether a point is inside a polygon (an "is-in" relation),
- what's the closest point (e.g. a business park) to a line (e.g. a road),
- how many polygons (e.g. cities) from one set contain a polygon from another set (e.g. parks),
- topological questions – such as "is there a hole in this polygon?".

These usually involve standard geometric operations as in Fig. 5.5. Surprisingly, there are few algorithms available that have been or perhaps even can be proved optimal for these types of operations. Computational Geometry remains a large and active field, though many geometric problems involving lines, areas and graphs now appear to be $\mathcal{NP}$-hard. As a very rough rule of thumb, you should assume that most 2D or higher dimension geometric problems are computationally "hard" (if you happen to get one that isn't, then you are lucky, and can sometimes publish a paper about your discovery). Luckily as data scientists we can leave most of this to Computer Science, which very helpfully implements

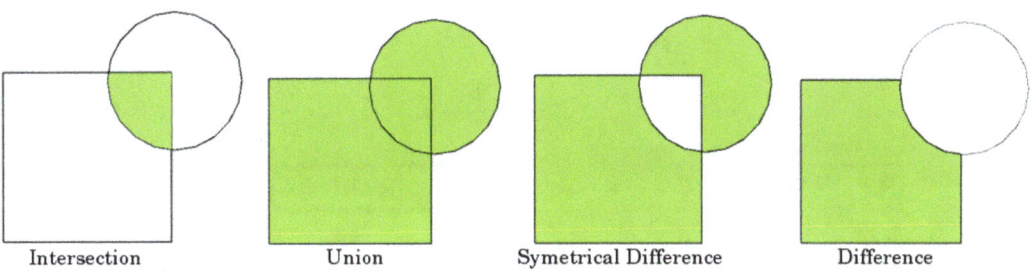

Fig. 5.5 Geometric operations

its best current solutions and makes them available for us as part of a GIS! In some cases, real-world objects might be represented by different vector entities at different scales. For example, a city is a polygon close up but a point in a satellite image.

The point/line/polygon ontology (together with various other types) and standard operations on its members is formalized by Open Geo-spatial Consortium (OGC) and by ISO19125 standards (also known as "simple features"), and used in most GIS systems. The standard calls individual spatial objects "features" and defines "layers" which group them together.

At the level of data ontology, there is a common distinction between vector and raster entities. Points, lines and polygons are vectors and are described by coordinates of their defining points (as in drawing programs). Raster entities are made of "pixels", usually square grids of individual data values. Example of raster entities include aerial drone photos (made of pixels, but with each image using a different coordinate system, including different scaling due to altitude changes); and also sampled data such as 2m grid terrain elevation data obtained from aerial lidar surveys (also sometimes having varying coordinate systems), or spatial surveys such as temperature, soil content, or traffic density in sampled locations across an area. Converting between raster coordinates (e.g. pixel row and column) is a complex task, but is automated by GIS tools. As with dates and times, don't try to do it yourself – use the tools!

Many vector and raster entities will carry additional, arbitrary properties. For example, we may assign street names to lines, or air pollution levels to polygon regions. A raster image will typically have location, scale, rotation and perhaps perspective, within the main vector space, e.g. information showing where an aerial photo is from.

5.3.3 Spatial Data Structures

We saw in Chapter 4 how a B+ tree can be used to implement a fast index into a one-dimensional database field. The B+ tree works by exploiting a property of one-dimensional data: that it can always be ordered. Even non-numerical, non-cardinal data fields such as text can be ordered by defining relations such as alphabetic or ASCII ordering. Real-valued data can be ordered by the positions of its values' locations in one-dimensional space. Hence B+ tree indexing on real-valued data is a simple form of spatial index, for single dimensioned data. Generalizing this idea to two (or more) dimensions is possible, though non-trivial. It can be done using a similar data structure to the B+ tree called an R-tree (Guttman, 1984), illustrated in Fig. 5.6. Here the two-dimensional space in (b) is clustered into rectangles (R), beginning with small ones representing spatial data entities of interest, then higher level ones acting as the index onto them. In the example, R8-R19 are real spatial entities, and R1-R7 form the index. The index is built and maintained by clustering nearby lower-level regions into higher level

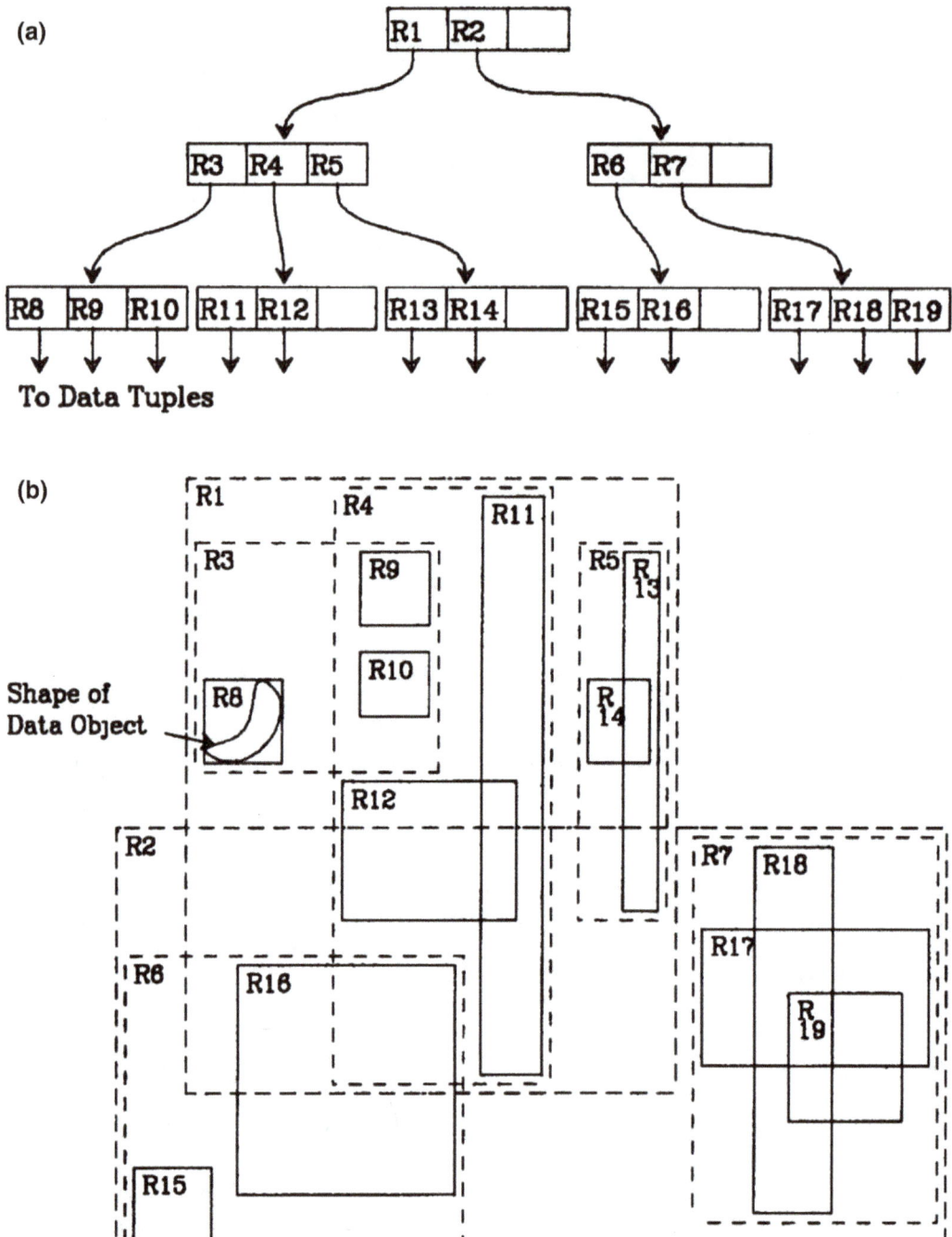

Fig. 5.6 R-tree structure. **a** Showing abstract data structure; **b** showing resulting partition of real 2D space

regions. Similarly to a B+ tree, this structure then allows one to search quickly for objects at particular locations. Unlike the B+ tree, the regions may appear in multiple parent regions, leading to the loss of the theoretical speed guarantees of B+ trees. Though in practice they usually work well and are used inside most GIS systems.

5.4 Implementations

Remember that not all transport tasks require spatial databases, and in many cases you can get by just by storing coordinates in a regular database. The spatial extensions are there for when you need to make lots of fast queries about inherently spatial relations such as quickly finding entities that are near or inside one another.

5.4.1 Spatial Files

Shapefiles are a storage format for the OGC ontology, used by most GIS systems. Contrary to their name, they are not single files but are small collections of related files, usually stored together in a directory. The main file has a *.shp* extension and stores the actual feature geometries. Other files that may appear along with it include *.dbf* (associated non-spatial properties data), *.shx* (indexing structure), and *.prj* (datum/projection information). For example, they are used in the Derbyshire data to store the locations of Bluetooth sensor sites as point entities.

GeoJSON and Well-Known Text (WKT) are alternative representation of the same ontology. Tools are available that will convert between these formats, in most GIS systems.

5.4.2 Spatial Data Sources

OpenStreetMap (*www.openstreetmap.org*) (OSM) was set up to counter problems with companies restricting access to map data, and is now a (just about) complete road map of the UK and many other countries, all in the public domain. OSM is usually the best place to start for map data. You can download and use for any purpose as much OSM data as you like free of charge. There are good free satnav apps such as *OsmAnd* which use this data, and which don't sell their users' personal GPS locations to any company. OSM data is stored in its own *.osm* format but tools are available to convert it to shapefiles and other common formats.

(If you ever notice any outdated or missing features in OSM, please fix them! It operates similarly to Wikipedia. If your country is lacking OSM data, please consider using one of its GPS-enabled phone apps to drive around yourself and collect it. This is often organized by groups of volunteers mapping new cities in "mapathon" days – perhaps you could lead one?)

Ordnance Survey OpenData makes some UK map data freely available, including terrain contour information, via its website,

www.ordnancesurvey.co.uk/business-and-government/products/opendata-products.html.

The UK Environment Agency makes much of its data freely available on the web, including very detailed lidar terrain scans up to 1m resolution, and environmental surveys such as noise, flooding and air quality levels, in some areas. (*www.gov.uk/check-local-environmental-data*). Its lidar data is used to power 3D "Earth view" type websites.

NASA (*data.nasa.gov*) satellite images are freely available, which are wrapped by "Earth-view" type websites. The original NASA data is much larger than that used on these sites, including images from multiple passes over the same sites. NASA data also includes non-visual spatial measurements from various environmental monitoring surveys.

5.4.3 Spatial Databases

The OGC standard is implemented, together with R-tree spatial index implementations, as an extension to SQL by several databases, including Postgres via its *PostGIS* extension (*postgis.net/documentation/*). These define new basic spatial types (ST). Other SQL extensions to databases may be available for high-level spatial tasks, for example the *pgrouting* extension for Postgres adds SQL commands to find shortest paths through road-like networks; and *pgpointcloud* adds tools for 3D point cloud data such as lidar scans from ground vehicles and aerial surveys.

5.4.4 Spatial Data Frames

We have previously discussed data frame structures used to manipulate data inside programming languages. A spatial data frame is simply a data frame which understands a database's representation of spatial data, which is usually stored in some non-human-readable binary code to enable fast spatial computations. Some spatial data frame libraries provide additional functions, such as automating projection conversions or assisting with map drawing.

5.5 Exercises

5.5.1 GPS Projections

A very common task is to convert co-ordinates, especially between ellipsoids and flat projections. This can be done with *pyproj* in Python, For example,

```
import pyproj
lat =   53.232350; lon = -1.422151
projSrc = pyproj.Proj(proj="latlon", ellps="WGS84", datum="WGS84")
projDst = pyproj.Proj(proj="utm", utm_zone="30U", ellps="clrk66")
(east_m, north_m)=pyproj.transform(projSrc, projDst,lon,lat) #non-ISO!
print(east_m, north_m)
```

5.5.2 PostGIS

PostGIS is a spatial database extension for Postgres. It installed in *itsleeds*, but (like all optional Postgres extensions) must be switched on in your database before it can be used. To do this, log onto your database with *psql* and give the command,

```
CREATE EXTENSION postgis;
```

Once this is done, you can create tables and data with new *geometry* types and Spatial Type (ST) functions such as the following, which can be used to store the locations of Derbyshire Country Council's Bluetooth sensor sites (as points), origin-destination routes between them (as straight lines connecting the origin and destination), and city boundaries in the area (as polygons, specifying their vertex coordinates in sequence, starting and ending at the same vertex). Points, lines and polygons are all stored in the same *geometry* type,

```
CREATE TABLE BluetoothSite (siteID text, geom geometry);
INSERT INTO BluetoothSite VALUES
   ('ID1003', 'POINT(0 -4)'),  ('ID9984', 'POINT(1 1)');
CREATE TABLE Route (name text, geom geometry);
INSERT INTO Route VALUES
   ('route1', 'LINESTRING(0 0,-1 1)'),
   ('route2', 'LINESTRING(0 0, 1 1)');
CREATE TABLE City (name text, geom geometry);
INSERT INTO City VALUES
   ('Chesterfield', 'POLYGON((0 0, 0 5, 5 5, 5 0, 0 0))');
```

Note the use of the *ST_AsText* function, which converts the data from the fast but non-human-readable binary code into human-readable text strings for display. You can see the raw binary code if you select spatial data without using this function,

```
SELECT siteID, ST_AsText(geom) FROM BluetoothSite;
SELECT siteID, geom FROM BluetoothSite;
```

Another common operation is to convert from spatial data to floating point numbers (especially if we want to compute vehicle speeds from location series) using a query such as,

```
SELECT ST_X(geom), ST_Y(geom) FROM BluetoothSite;
```

And queries about spatial relations between entities such as asking for all the Bluetooth sites located in the boundary of Chesterfield,

```
SELECT BluetoothSite.siteID
FROM BluetoothSite, City
WHERE ST_Contains(City.geom, BluetoothSite.geom)
AND City.name='Chesterfield';
```

(The coordinates in the above examples use integers which might, for example represent coordinates in meters referenced to some origin. For real data we would use either GNSS coordinates or some standard projection. The choice of which to use is important for computation speed: often you will want to work with projected coordinates stored in the database, to avoid having to re-project them every time they are used to calculate or draw anything.)

5.5.3 GeoPandas

Python's spatial data frame library is called GeoPandas. GeoPandas extends Pandas' DataFrames to represent and manipulate OGC data in its GeoDataFrames. It can connect to a PostGIS enabled Postgres database as standard Pandas can be connected to a standard Postgres. GeoPandas can read data into DataFrames from a spatial database like this,

```
import geopandas as gpd
import psycopg2
con = psycopg2.connect(database='mydatabasename', user='root')
```

```
sql = "SELECT * FROM BluetoothSite;"
df = gpd.GeoDataFrame.from_postgis(sql,con,geom_col='geom')
print(df['geom'][0].coords.xy)    #get coordinates as numbers
for index, row in df.iterrows(): #loop over rows
    print(row)
```

GeoPandas can also load shapefiles directly into GeoDataFrames like this,

```
df=gpd.GeoDataFrame.from_file('data/dcc/examples/ BluetoothUnits.shp')
```

We can insert some of the contents of a GeoDataFrame into a database table by iterating over it and using SQL commands,

```
for i in range(0,df.shape[0]):
  sql="INSERT INTO BluetoothSite VALUES \
    ('%s', '%s');"%(df.iloc[i].Site,df.iloc[i].geometry)
  cur.execute(sql)
  con.commit()
```

As Pandas replicates much SQL functionality inside Python, so GeoPandas replicates much PostGIS functionality inside Python. This includes Python commands for performing spatial operations. Also like Pandas, it contains tools to assist with data munging for spatial data (such as shapefiles). There are also functions to assist with projection, such as the following to specify the initial coordinate reference system (CRS) then convert all geometry in a GeoDataFrame to a new CRS (in this case, British National Grid, also known as EPSG:27700) at once, which automates many calls to *pyproj*,

```
df.crs = {'init': 'epsg:4326', 'no_defs': True} #initial CRS
df = df.to_crs(epsg=27700)    #CRS conversion
```

5.5.4 QGIS Road Maps

Start QGIS. (From the Desktop: *Applications* → *Education* → *QGIS Desktop*).

Open the shapefile, *data/dcc/examples/BluetoothUnits.shp* in QGIS to see the Bluetooth sensor locations in Derbyshire. (*Layers*→*Add layer* → *Add vector layer*. Select the *.shp* file for *dataset*.)

To include a map of Chesterfield in the background, from an OpenStreetMap server:

- Menu *Vector* → *OpenStreetMap* → *Load data* with "from layer" option will connect to the OSM server and download data to match the sensor layer's area. (Give it a filename.)
- Menu *Vector* → *OpenStreetMap* →*Import topology from an XML file* to convert the OSM XML data into a QGIS database file. (Load from the filename.)
- Menu *Vector* → *OpenStreetMap* → *Export topology to Spatialite* to load the database file (filename+'.db') into a QGIS layer. Select the *polyline* option to display roads as lines.
- Change the ordering of the layers by dragging and dropping to show the Bluetooth sites on top of the map.

5.5 Exercises

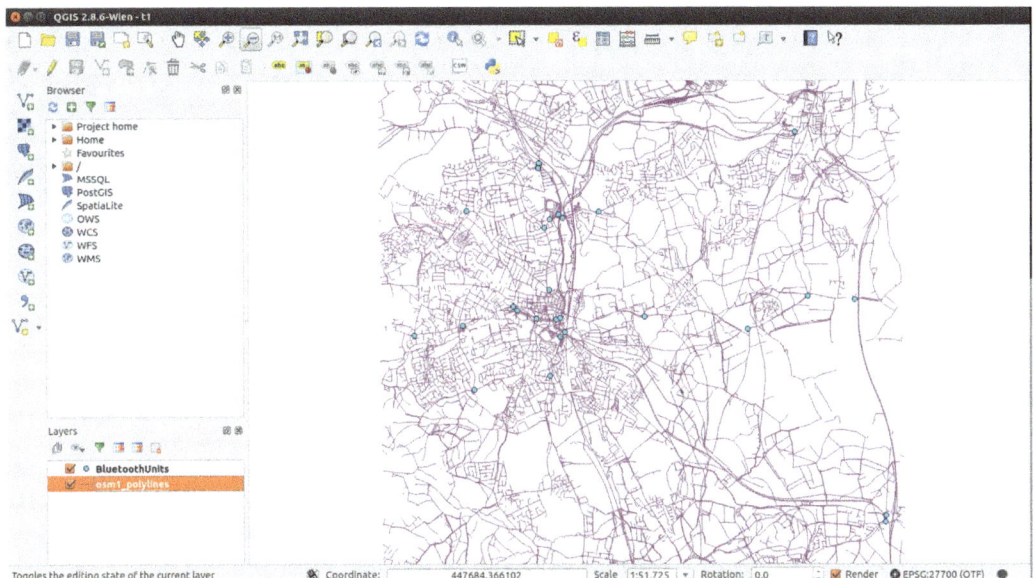

Try plotting other sensor locations on the map too. Play with QGIS to see what it can do. (Full documentation: *www.qgis.org*).

5.5.5 Plotting Open Street Map (OSM) Roads

The Docker image contains a Shapefile version of Derbyshire OSM data, *dcc.osm.shp* in its *data* folder. *Tasks:*

- Load the Derbyshire Shapefile into GeoPandas.
- Convert it from GPS coordinates to National Grid (meters) projection (e.g. with PyProj or GeoPandas).
- Create Postgres tables for it and store.
- Plot the major road types in different colors (use *df_roads['highway']*).

Hints:

```
#importing roads shapefile to database
fn_osm_shp = "data/dcc.osm.shp/lines.shp"
df_roads = gpd.GeoDataFrame.from_file(fn_osm_shp)
df_roads = df_roads.to_crs({'init': 'epsg:27700'})
for index, row in df_roads.iterrows():
  sql="INSERT INTO Road VALUES ('%s', '%s', '%s');" \
  %(row.name, row.geometry, row.highway )
  cur.execute(sql)
con.commit()
#road plotting
sql = "SELECT * FROM Road;"
df_roads =  gpd.GeoDataFrame.from_postgis( \
  sql,con,geom_col='geom') #
print(df_roads)
for index, row in df_roads.iterrows():
```

```
    (xs,ys) = row['geom'].coords.xy
    color='y'     #road colour by type
    if row['highway']=="motorway":
        color = 'b'
    if row['highway']=="trunk":
        color = 'g'
 plot(xs, ys, color)
```

5.5.6 Obtaining OSM Data

To obtain and convert OSM data to Shapefiles yourself, from the command line, download OSM data for Chesterfield and convert it to standard ShapeFile format. For small maps this can be done with *wget*,

```
$ wget http://api.openstreetmap.org/api/0.6/map?bbox=\
-1.4563,53.2478,-1.4011,53.2767 -O ~/dcc.osm
```

(*-O* specifies where to save the *o*utput.) The URL can also be entered into a web browser if the *wget* command is not available.

For larger maps (which place more load on OSM's servers), *wget* requests may be rejected by the server, and the OSM project asks users instead to use its alternative "Overpass API" run on volunteer servers such as *http://overpass-api.de/query_form.html*. This will return your data if you paste in XML queries such as,

```
<union>
  <bbox-query s="53.1567" w="-1.5065" n="53.2835" e="-1.2971"/>
  <recurse type="up"/>
</union>
<print mode="meta"/>
```

This is the query used to download the Docker data. You may need to manually rename the downloaded file to *~/data/dcc.osm*.

The OSM file can then be converted to a ShapeFile with,

```
$ ogr2ogr --config OSM_USE_CUSTOM_INDEXING \
NO -skipfailures -f "ESRI Shapefile" data/dcc.osm.shp/ data/dcc.osm
```

5.5.7 Bluetooth Traffic Sensor Sites

Tasks:

- Load the real Bluetooth site locations into the database, using the spatial *POINT* style. (In *data/dcc/web_bluetooth_sites.csv* the locations arrive as single strings of 12 digits each. The first six and last six are the two coordinates.)
- Load them into GeoPandas from the database.
- Plot them on top of the OSM road map above as blue circles.

5.5 Exercises

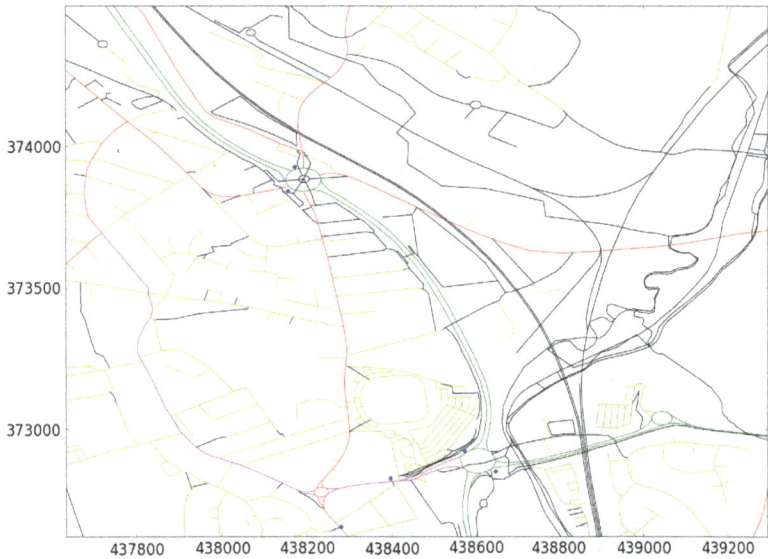

Fig. 5.7 Plotting Open Street Map roads and Derbyshire Bluetooth sensor sites together

Hints:

```
#bluetooth site plotting
sql = "SELECT ST_x(geom), ST_y(geom), site, geom \
  FROM BluetoothSite;"
df_sites = gpd.GeoDataFrame.from_postgis( \
  sql,con,geom_col='geom')
plot( df_sites['st_x'] , df_sites['st_y'], 'bo' )
```

A full program example is provided in the Docker image – but try to write it yourself first. Your final plot should look like Fig. 5.7.

5.6 Further Reading

- Guttman A (1984) R-Trees: a dynamic index structure for spatial searching. Proceedings of the 1984 ACM SIGMOD international conference on Management of data – SIGMOD '84 (Original R-tree source)
- DeMers MN (2009) GIS for dummies. Wiley. (Good beginners' guide to GIS general concepts)
- Noureldin A, Tashfeen BK, Jacques G (2012) Fundamentals of inertial navigation, satellite-based positioning and their integration. Springer Science & Business Media. (Very comprehensive reference guide to GPS and IMU systems and processing)
- Shekhar S, Chawla S, Ravada S, Fetterer A, Liu X, Lu CT (1999) Spatial databases-accomplishments and research needs. IEEE Trans Knowl Data Eng 11(1): 45–55
- Links to many interesting papers and reviews on related topics: http://gis.usc.edu/msp-resources/articles-blogs/the-role-of-spatial-database-in-gis/
- geopandas.org (full GeoPandas documentation)
- The Python GIS cookbook: https://pcjericks.github.io/py-gdalogr-cookbook/ (full of practical, runnable code examples of most tools, usually up to date)

(from www.xkcd.com)

Appendix: Inside GeoPandas

GeoPandas wraps a large stack of Python tools, making them easy to use but hiding the detail of what they do and how they relate to the systems we have seen in previous chapters. For many transport applications GeoPandas is sufficient, but sometimes you may need to go inside and use the lower level tools directly. This section gives a short overview of the stack.

GDAL/OGR is an OGC implementation library with wrappers available for many programming languages, including Python. It represents the ontology as classes within the programming language rather than in a file or database. OGR is the part that handles OGC vectors, while GDAL handles rasters. To use it with a particular language such as Python, you need to install both the core library and a wrapper for that language.

Example code using GDAL/OGR without GeoPandas is given for Derbyshire map plotting. It represents OGC features directly as Python objects (rather than GeoDataFrames), having geometry and other fields as members.

If you need to use more advanced spatial operations, similar to those available in spatial databases such as intersections and overlaps, then the Python-specific library Shapely provides many of those functions. (In fact Shapely is used in the implementation inside some of the spatial databases). Shapely works with WKT formatted data so you need to convert it then load it back in from WKT format like this,

```
wkt = geometry.ExportToWkt() #Shapely works with wkt format,
convert shape = shapely.wkt.loads(wkt)
```

If you need to import or export more exotic file formats, the Fiona library does this and is available for many languages including Python.

A typical Python GIS application, such as processing the Derbyshire data, might have a structure such as shown in Fig. 5.8.

In this architecture, shapefiles are used as the main input/output formats, so other formats are converted to them, including OpenStreetMap's own data format *.osm,* and displayable *.png* output images. The OGC data in the shapefile appears in Python as an "OGR Datasource" which contains layers of features of geometry and fields. Geometry holds spatial data defining entities and fields hold non-spatial data about those entities (such as their names, observed traffic counts etc.) The application might interface to a spatial database such as Postgres/PostGIS via an SQL query API; and/or it might convert some of the data to WKT format and run Shapely functions on it.

Here is an example of using OGR directly to plot the map of Derbyshire,

```
import ogr
from matplotlib.pyplot import *
ds = ogr.Open("/headless/data/dcc.osm.shp/lines.shp")
datasource layer = ds.GetLayer(0) nameList = []
for feature in layer:
    col="y"     #change colour if its an interesting road type
    highwayType = feature.GetField("highway")
    if highwayType != None:
        col="k"
    if highwayType=="trunk":
        col="g"
    name = feature.GetField("name")
    nameList.append(name)
    #get the features set of point locations (a wiggly line)
```

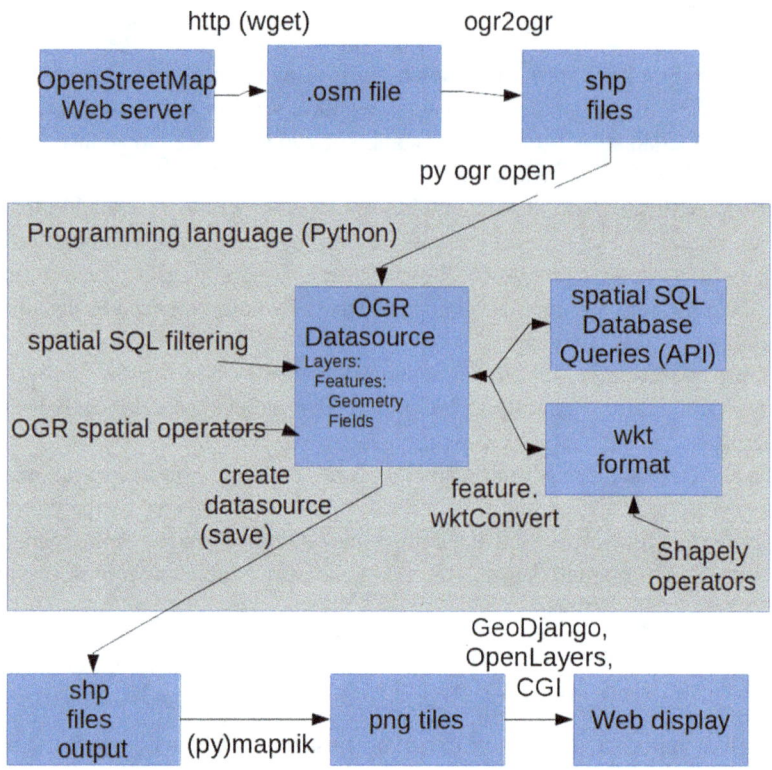

Fig. 5.8 OGR-based application architecture

```
geomRef=feature.GetGeometryRef()
x=[geomRef.GetX(i) for i in range(geomRef.GetPointCount())]
y=[geomRef.GetY(i) for i in range(geomRef.GetPointCount())]
plot(x,y, col)
```

And an example of extracting text names of objects from the Bluetooth shapefile,

```
import ogr
from matplotlib.pyplot import *
ds = ogr.Open("/headless/data/dcc/examples/BluetoothUnits.shp")
layer = ds.GetLayer(0) #shapefiles may have multiple layers
ldefn = layer.GetLayerDefn() #loop over each feature
for n in range(ldefn.GetFieldCount()):
    featurename = ldefn.GetFieldDefn(n).name
    print(featurename)
for feature in layer: #loop over each object in the layer
    location_description = feature.GetField("Location")
print(location_description)
```

Bayesian Inference

6

Bayesian inference, otherwise known as "probability theory", is the theory of how to combine uncertain information from multiple sources to make optimal decisions under uncertainty. These sources include empirical data and also other beliefs, called *priors*, which may come from previous experiments, theory, and subjective estimates. Bayesian theory makes probabilistic *inferences* which are complete probability distributions over our beliefs about unobserved variables of interest, including generative parameters of models assumed to cause to data, as well as unobserved variables which are caused by the observed data. Bayesian inference is computationally hard, so typically works with approximate calculations on large compute systems. Bayesian inference is provably (Bernardo and Smith 2001) the *only* system able combine beliefs to make decisions consistently and optimally.

6.1 Bayesian Inference Versus "Statistics"

Bayesian inference is sometimes called "Bayesian Statistics" though this can be viewed as something of a misnomer. A *statistic* is a function of data which is used to estimate or describe something, usually by reducing a set of data to a single number. Bayesians do not estimate values of unobserved variables with single numbers – instead they infer the complete probability distribution over *all* possible values of the variables. Before Bayesian inference became computationally practical, statistics were used heavily in the field known as "Frequentist Statistics", "Classical Statistics", or just "Statistics" which is still taught in some schools and universities. Statistics as a field also made use of concepts including null hypotheses and *p*-values which have been superseded by related Bayesian concepts. As a community, Statistics typically divided into two professional roles, with mathematicians inventing new statistics and procedures for their use, and users learning how to apply these pre-built tools to their problems. In contrast, Bayesian inference professionals divide into users who build their own mathematical models of assumed generative processes for each data set, and programmers who provide reusable computational approximation systems to perform inference on these models. Statistics generally assumes that its objective is to concisely describe the data ("how many vehicles are there?"), while Bayesian inference considers the wider objective of prescribing practical actions based on the data ("how should we program the traffic lights?"). While results from multiple experiments from Classical Statistics can be combined by human specialists in "meta-analysis", the results of Bayesian inferences come in the right format to immediately form priors for future inferences. For example, observations taken 10 times per second by an autonomous vehicle can be viewed as a sequence of "experiments" and Bayesian inference easily enables them to be combined at each moment to give the most up to date view of the world. While it is possible to do Data Science with Statistics, Bayesian inference's emphases on computation, mathematical understanding, action selection, and fusing uncertain data from multiple ("variety"), messy ("veracity") and continually updated ("velocity") sources make it the more natural mathematical tool for Data Science.

6.2 Motorway Journey Times

To illustrate the Bayesian approach and compare it to a Classical Statistics analysis, suppose we are interested in deciding when to leave Sheffield to drive to work in Leeds in the morning. A Frequentist statistician might go out and collect N observations of travel times, x, throughout the morning over several weeks, then compute a sample mean statistic and standard deviation statistics from journey times, for each departure time window. She would probably create some null hypothesis, such as "journey times are independent of departure time", and proceed to test and reject this hypothesis by showing that there is a low probability (p-value) of it giving rise to the observed, or worse, data. Then she will show that her sample mean statistic,

$$\bar{x} = \frac{1}{N} \sum_{n=1}^{N} x,$$

and sample variance statistic,

$$s = \frac{1}{N-1} \sum_{n=1}^{N} (\bar{x} - x_n)^2,$$

converge, in the limit of infinite data, to the true population mean and variance. She might use further statistics to give confidences about how far these have converged and how likely it is that the observed statistic values have arisen by chance. If you ask her why the first estimator contains $1/N$ and the other contains $1/(N-1)$, she probably doesn't know, and will either try to talk about "degrees of freedom" or mention that it was proved in a textbook by someone else a long time ago. Similarly, the statistics used for the p-value and other confidence measures will come from an old book or tables, and are not the kind of thing that a working user will worry about. If this book does not contain a suitable recipe for the problem, she will change the problem or assumptions or work on a different problem that does have a recipe. The analysis is based only on observed data, and it does not consider the interaction of the reported results with the desired action of actually driving to work. It treats the world as objective – there is some true mean and variance and the job of statistics is to estimate and report them, regardless of how they will affect your driving.

In contrast, a Bayesian may think differently. Before thinking about data he will consider what the study is for. He will write down the set of possible actions he can take in the real world: in this case, the set of times of the morning he could depart for work. He might include other types of actions too, such as staying home and dialing into work via Ekiga or Skype. Next he considers the utility of each of these actions in each possible world that may occur. Leaving late has a dollar value of the extra minutes allowed to stay in bed. If traffic is good and he arrives on time, there is no dollar penalty. But if he has an important meeting at 9am then there is a dollar loss for being late. Staying home and dialing in would get the meeting done but might not establish as good a relationship with the client, so scores some dollar utility but not as much as being there in person. After thinking about action and utility, he then considers probabilities, but not yet data. What does he already believe about journeys

6.2 Motorway Journey Times

on this motorway? Probably he has driven the route many times before without explicitly collecting data, and also heard reports from friends and on the radio about similar journeys at different times. All these friends and radio stations have different reliabilities – some friends exaggerate their stories while others are known to be trustworthy – and these probabilities of accuracy should be taken into account. Before the Bayesian collects any new data, he can combine all of this prior information into a belief about journey time as a function of departure time. He will assume some generative model M_1 based on it, such as journey times being Gamma distributed (with parameters k and θ together specifying the distribution similarly to mean and variance together),

$$p(x) = \frac{1}{\Gamma(k)\theta^k} x^{k-1} e^{-\frac{x}{\theta}}.$$

The choice of this model might be based not only on assumptions about journey times in general, but also on the Bayesian's computational needs – he will often select an exponential form like this to make the maths and computations easier later on, even if it is not a perfect model of reality. In many cases, the Bayesian might not need to collect any new data at all, because the combination of prior beliefs, reports, and credibilities may be enough to optimize the choice of action. When they are not sufficient, the Bayesian can compute exactly *how* insufficient they are, and work out how much additional data is needed to collect to make them sufficient when combined with all the prior knowledge. If the Bayesian gets into an argument with another Bayesian about the relevance of the Gamma model, he challenges her to provide an alternative model, M_2. Perhaps she proposes a mixture of Gamma distributions intended to model two regimes of a phase change in traffic. Neither Bayesian needs to care about how theoretically realistic these models or their assumptions are. Instead, they can both be tested against the data and will result in probabilities that the models are correct, $P(M_1|D)$ and $P(M_2|D)$. They might then both use the best model to plan their journeys – or even a mixture of the two, via,

$$\hat{a} = \arg_a \max \sum_s U(a,s) P(s|D,M),$$

where a are the possible actions, $\hat{a}$ is the optimal action, s are states of the world, D is the data, U is the utility function, and M is the model. This will probably require the use of a large computer running approximate numerical calculations. The two Bayesians have not used any "statistics" to estimate anything – rather they talk directly about probabilities, models, parameters, actions and utilities. They have not claimed any objective knowledge about the world – and might each use and "believe" in different models that best optimize their own journeys, work, and dollar-value circumstances.

Frequentists may argue that it is possible to do Data Science without Bayesian theory, for example by applying "old fashioned" classical statistical methods to data in databases. Criticism from the opposite side – the young and fashionable "big data" movement – argues that as data sets become "big" then neither Frequentist nor Bayesian analysis is longer needed at all, because previously unobservable populations may become fully observable from, rather than sampled by, these data sets ("N = all"). However, Bayesian theory's emphasis on practical utility, approximations, joining of information from multiple sources, working with noise and messiness, and computational hackery are a close cultural fit to the Data Science community in general. And unless you are certain that you really have a perfectly noiseless collection of every single member of the population there will still be some uncertainty to deal with using probabilities. We don't have access to "big data" from the future either, so any inferences about predictions will still involve probabilities. In Derbyshire we might have access to "a lot" of data but as a rural area we can't afford to cover every part of the network with sensors, so we have to make inferences about non-covered areas.

6.3 Bayesian Inference

6.3.1 Bayes' Theorem

Bayes' "theorem" is actually more of a "lemma"[1] as it follows trivially from two definitions in probability theory: marginalization,

$$P(A) = \sum_B P(A, B),$$

and conditioning,

$$P(A|B) = \frac{P(A, B)}{P(B)},$$

giving Bayes' theorem,

$$P(B|A) = \frac{P(A|B)P(B)}{P(A)}.$$

Proof: exercise (easy!).

The genius of the theorem comes from the conception of probability that enables the definitions in the first place. This was widely debated in the time leading up to the theorem, by a large community of philosophers and mathematicians, largely due to the recent inventions and/or legalizations of gambling games including dice, cards, insurance, and investment banking.

6.3.2 Legal Inference: A Pedestrian Hit-and-Run Incident

Suppose you have been called up for English jury service in a manslaughter trial. A hit-and-run incident was captured on a CCTV camera, where the driver hit and killed a pedestrian then drove away. The CCTV shows a blurry image of the driver escaping from the scene in their vehicle. After some image processing, ANPR is able to recognize the licence plate, but not the make or model of the vehicle or any of the driver's features. A Bluetooth sensor near the crime scene also picked up the driver's mobile phone signature matching the CCTV footage. However, both the ANPR and Bluetooth sensing are noisy and error prone. An expert statistical witness in the court (who has a 100% reliability track record) states that the odds of this ANPR matching a random person's vehicle by chance are 1 in 1000, and the odds of this Bluetooth sensor matching a random person's phone are 1 in 10,000. This data and testimony are the only evidence presented by the prosecution. Under English law you are required to vote for "guilty" or "not guilty", where you should choose "guilty" if you believe the suspect to have committed the crime "beyond all reasonable doubt". The meaning of "reasonable doubt" is deliberately left vague in English law and is for you to decide yourself according to your own ethics and knowledge.

---

[1] Bayes' theorem is also almost certainty misnamed in its discoverer too. The leading history of this reads like a detective novel itself (Stigler, Stephen M. *Who discovered Bayes' theorem?*, The American Statistician 37.4a (1983):290–296) and concludes that the discovery is most probably due to Nicholas Saunderson, who grew up in a village 25 miles from ITS Leeds, and – amazingly – performed all of his mathematics research after going blind, using a tactile abacus of his own invention.

- How will you vote?

To model this scenario with Bayes' theorem, let event M mean that the suspect is the manslaughterer, and event S be the observed sensor data (ANPR plus Bluetooth). The expert witness has told us that for a random person,

$$P(S) = 1/1,000 \times 1/10,000 = 1/10,000,000,$$

and we will assume that the sensors will give S if the suspect really is the manslaughterer,

$$P(S|M) = 1.$$

What is the prior probability that the suspect is the manslaughterer, before we take account the evidence presented to us in court? We do not know anything about how this person was caught by the police, for example whether they chased the car at the crime scene and arrested the suspect there, or whether they accessed their database of licence plates and mobile phone data to find the suspect. In the absence of any of this type of evidence, we must assume that the prior probability that the suspect is the manslaughterer is equal to the prior probability that any arbitrary person is the manslaughterer. Assuming that the manslaughterer is at least a UK driving licence holder, there are about 40 million such people in the UK, so the prior belief that any one of them happens to be the manslaughterer, $P(M)$, is 1/40,000,000.

Combining these probabilities with Bayes' rule gives,

$$P(M|S) = \frac{P(S|M)P(M)}{P(S)} = \frac{(1)(1/40,000,000)}{(1/10,000,000)} = \frac{1}{4}.$$

If you voted to execute or imprison this suspect then you have done so with a 3/4 probability that the suspect is innocent!

This shows a fundamental problem with using data-mining to generate hypotheses, known as the "prosecutor's fallacy". For most types of hypotheses that can be formed based on large data sets, the large data also specifies a large number of variations upon those hypotheses. This means that the prior probability of any one being true is very low, and the evidence presented to overcome it must be equally powerful.[2]

6.3.3 Priors and Posteriors

The fundamental step in any Bayesian inference is the use of Bayes' theorem to update a prior belief such as $P(M)$ into a more refined posterior belief such as $P(M|S)$ by using some data S and a model of how the data is generated, $P(S|M)$. Bayesians view the theorem as a computation update rather than just an equality,

$$P(B|A) := \frac{P(A|B)P(B)}{P(A)},$$

where := in Computer Science means "set the value of the term on the left equal to the computed result of the calculation on the right".

---

[2] If this sounds far-fetched, a statistically similar miscarriage of justice actually happened in 1999, sentencing Sally Clerk to two life sentences for what was later proven in *(R. vs. Clerk. 2003)*, using Bayesian analysis, to be two coincidental cot deaths.

Typically this update is computationally hard, i.e. $\mathcal{NP}$-hard, which means that even with supercomputers or cloud clusters we still can't always perform it exactly. The reason for this is that the $P(A)$ term is hard to compute, as it must sum the probability of A under every possible counter-factual value of B. B often has a huge potential state space, as it could be a complex state made from a combination of parts, leading to an exponential number of possible combinations. And/or it could be continuous valued, requiring integration,

$$P(B|A) := \frac{P(A|B)P(B)}{\int P(A|B')P(B')dB'}.$$

Instead of calculating this exactly, most working Bayesians use standard numerical approximation tools to give approximate answers. The details of these tools are beyond the scope of this book, but we will say that one such method is called MCMC (Markov Chain Monte Carlo) and operates by drawing millions of random samples in place of performing exact integrations. Software tools exist such as PyMC3 which perform MCMC in response to the user specifying a Bayesian model, without the user needing to know exactly how they work inside.[3]

6.3.4 Road User Tracking

Bayes' theorem forms the foundation of tracking algorithms for vehicles and other road users, from CCTV, GNSS, or other sensors. For example, the previous chapter discussed the need to combine GNSS data with historical readings and other data sources to obtain accurate location estimates over time for autonomous driving; and the M25 motorway example in Chap. 1 involves tracking vehicles over long distances between ANPR cameras. A "track" is a temporal sequence of inferred locations. Bayesian tracking will typically have "filtering" forms to introduce each time t's new observation data O_t, such as,

$$P(X_t|O_{1:t}) = \frac{P(X_t|O_t)P(X_t|O_{1:t-1})}{Z},$$

where X is the state of the vehicle (including position, heading, velocity and possibly higher order derivatives such as acceleration), Z is chosen to normalize the sum of the probabilities to unity, and the prior $P(X_t|O_{1:t-1})$ decomposes as the sum of factors for previous state beliefs given by the historical data, $P(X_{t-1}|O_{1:t-1})$, and for a physical motion model, $P(X_t|X_{t-1})$, to give,

$$P(X_t|O_{1:t}) = \frac{P(X_t|O_t)\int P(X_t|X_{t-1})P(X_{t-1}|O_{1:t-1})dX_{t-1}}{Z}.$$

In general this must be solved by MCMC methods, usually "particle filters" which are special cases of MCMC specialized for temporal filtering. In (rare) cases where all terms are linear and Gaussian, there is an analytic solution called the "Kalman Filter".

6.4 Bayesian Networks

A *Bayesian Network* represents the causal relationships between the variables as a graph. Formally, the terms $P(X_i|\{X_i^{(1)}, \ldots, X_i^{(N_i)}\})$ which together define the joint distribution,

---

[3]The other main standard approximations are "loopy belief propagation", "Variational Bayes", and assuming various distributions to make exact inference tractable.

6.4 Bayesian Networks

$$P(X_1, X_2, \ldots, X_M) = \prod_i P(X_i|\{X_i^{(1)}, \ldots, X_i^{(N_i)}\}),$$

where $X_i^{(j)} \in \{X_k\}_k$, are represented by graphical nodes labeled X_i with directed links connecting their parents $X_i^{(j)}$ to them.

6.4.1 Bayesian Network for Traffic Lights

To illustrate a simple Bayesian network: suppose we are concerned that a two-way traffic light at a roadworks site is behaving unfairly, giving priority to one direction more often than the other. To simplify the problem, we will assume the light is either red or green at each point in time, from the direction we observe it.

Assume the state of the light is generated by a parameter θ, such that $P(green) = \theta$ and $P(red) = 1 - \theta$. Our prior belief about the value of θ contains no information except that its value is between 0 and 1. This is called a "flat", or "uniform", prior. We observe N data points, $color_{1:N}$, and want to compute our posterior belief $P(\theta|color_{1:N})$. We will model our belief distributions over θ using Beta distributions, which are quite like Gaussians but squashed into the range [0, 1], because we know that θ is itself a probability so must lie in this range. You don't need to know the Beta equations here because the Beta model will be computed by a computer library, but they are defined by two parameters (α, β) which together control their mean and variance. Some examples of Beta distributions with different values of these parameters – including the flat prior case $(\alpha, \beta) = (1, 1)$ as well as posterior-type curves – are shown in Fig. 6.1.

The simple Bayesian network model displayed as a graph in Fig. 6.2, specifies that the bias of the traffic light, θ, causes the values of the observed colors, so that the joint probability is,

$$P(\theta, color) = \prod_i P(X_i|\{X_i^{(1)}, \ldots, X_i^{(N_i)}\}) = P(\theta)P(color|\theta).$$

Using Bayesian network software, approximate probabilities of variables in the network can be computed automatically – remembering that inference is computationally hard so only approximations

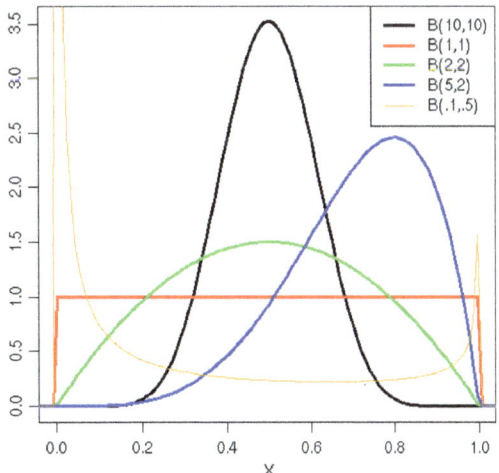

Fig. 6.1 Examples of Beta probability distributions

Fig. 6.2 Traffic light Bayesian network

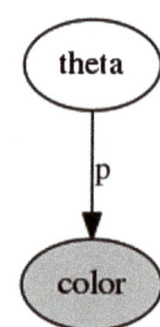

Fig. 6.3 Inferred posterior belief for parameter θ

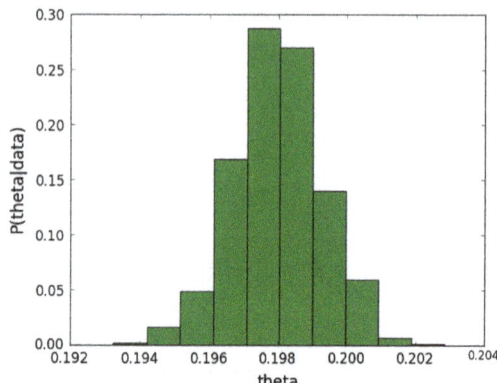

are computed. Figure 6.3 shows the inferred posterior for θ after presenting some data observations. From the shape of this posterior, we are now very certain that the value of θ lies in the region around 0.198. A frequentist analysis would probably use the sample mean of the colour observations, together with further statistics such as z-scores to give confidence that this sample mean is a close estimator of the true θ. In contrast, the Bayesian analysis shows exactly what our belief about $P(\theta)$ actually is. You can read whatever you like off this histogram to report statements such as, "we are 97% certain that θ lies between 0.195 and 0.201".

6.4.2 Bayesian Network for Traffic Accidents

We will next consider a more complex model to see how Bayesian networks generalize easily where simple Frequentist statistics do not.

Suppose we have observed the number of accidents on a road per month for the last 110 months, which are,

[4, 5, 4, 0, 1, 4, 3, 4, 0, 6, 3, 3, 4, 0, 2, 6, 3, 3, 5, 4, 5, 3, 1, 4, 4, 1, 5, 5, 3, 4, 2, 5, 2, 2, 3, 4, 2, 1, 3, 2, 2, 1, 1, 1, 1, 3, 0, 0, 1, 0, 1, 1, 0, 0, 3, 1, 0, 3, 2, 2, 0, 1, 1, 1,0, 1, 0, 1, 0, 0, 0, 2, 1, 0, 0, 0, 1, 1, 0, 3, 3, 1, 1, 2, 1, 1, 1, 2, 4, 2, 0, 0, 1, 4, 0, 0, 0, 1, 0, 0, 0, 0, 0, 1, 0, 0, 1, 0, 1].

Plotting this data suggests a theory that something changed part way through the time series which reduced the accident rate. Assume that under any fixed environmental state, the number of accidents, k, is Poisson distributed (i.e. the state has one parameter, λ, which makes accidents more or less probable; and the accidents within each month don't affect each other),

6.4 Bayesian Networks

Fig. 6.4 Bayesian network for accidents model

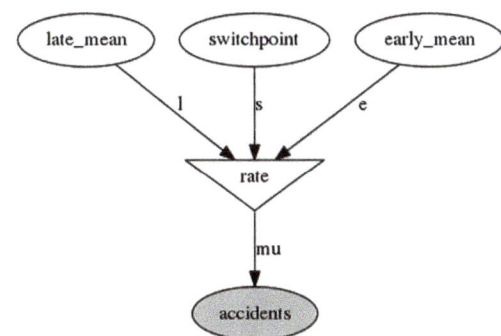

$$P(k) = \frac{\lambda^k e^{-k}}{k!}.$$

We might construct a model that assumes there were two distinct states of the environment, with one higher-accident state prevailing until some point in time, then switching to the lower-accident state after that. If we can infer the parameters of this model, then we can read off the most probably times when the switching occurred, which will help us to look at historical records about the road infrastructure and determine what the cause might be. Perhaps a new road sign was installed then, or some new construction project altered the commuting patterns.

The Bayesian Network shown in Fig. 6.4 implements this model, and is equivalent to,

$$P(\{X_i\}) = \prod_i P(X_i | \{X_i^{(1)}, \ldots, X_i^{(N_i)}\})$$

$$= P(late\_mean) \times P(early\_mean) \times P(switchpoint)$$
$$\times P(rate | late\_mean, early\_mean, switchpoint) \times P(accidents | rate),$$

where the *early_mean* and *late_mean* are the Poisson mean parameters (λ), and we assume Exponential Distribution beliefs over them. We also assign a uniform prior over the time where the switch-over occurs (switchpoint).

6.4.3 Reporting

Frequentists usually report results about individual variables. The Bayesian posterior graphs in the above example are similarly about individual variables. However, Bayesians are not limited to this type of reporting.

Sometimes we might be interested in the *joint* distribution of variables. For example we might plot the joint posterior of all three (*late_mean, switchpoint, early_mean*) in a 3D space as in Fig. 6.5.

In some cases, considering the joint can give a very different picture of what is going on than looking at individual variable posteriors. In the example above, there are several different regimes which are all good fits to the data, but require all the variables to be set together to get the fit. Taking the best value of each variable individually could result in a mixing-up of aspects of each regime, and a worse overall fit.

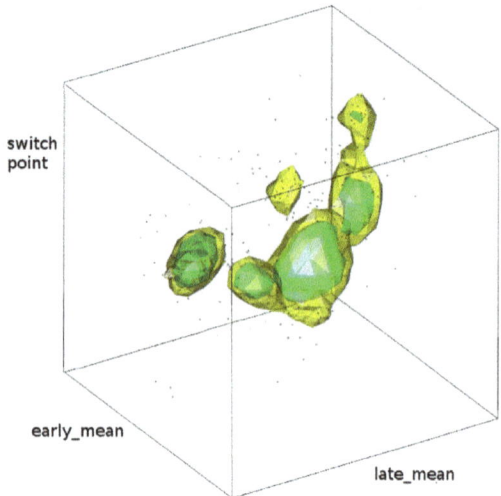

Fig. 6.5 Example of a complex joint posterior distribution. Yellow and green show thresholded medium and high probability densities

As a general principle, Bayesians do not even report posteriors at all. Rather, they use the posterior, computationally, to inform the selection of some real world optimal action, $\hat{a}$, via,

$$\hat{a} = \arg_a \max \sum_s U(a,s) P(s|D).$$

Academic research is – in theory – an unusual, special case for Bayesians, in that writing a research paper usually does require making a claim about some "objective fact" found from the data, rather than taking a physical action. This forms a special case whose utility function is simply assigns some fixed positive dollar value for the act of reporting the truth, and zero otherwise. As many successful transport consultants know, the cash utility of their commercial consultation reports may vary with many other political factors, including "what the client wants to hear", "what the client's boss wants to hear", and most importantly, "what findings will generate the most future consulting projects".[4]

6.4.4 Car Insurance

Car insurance companies now routinely use Bayesian Network models to understand accident probabilities and factors, and to set premiums. These models can become very complex and seek to capture interactions between many relevant factors, such as the example in Fig. 6.6.

---

[4] Academic paper publications are not completely free from these problems in practice, as their values are also measured in cash values by modern university managers. Bayesian data scientists make up a large part of the "fake science" movement which seeks to expose scientific malpractice, such as reporting of false results or statistical interpretations for financial gain. They argue that while Science's theoretical utility lies in reporting the truth, individual researchers are more strongly motivated by the needs to publish and advance their careers under current research management models. See *www.callingbullshit.org*.

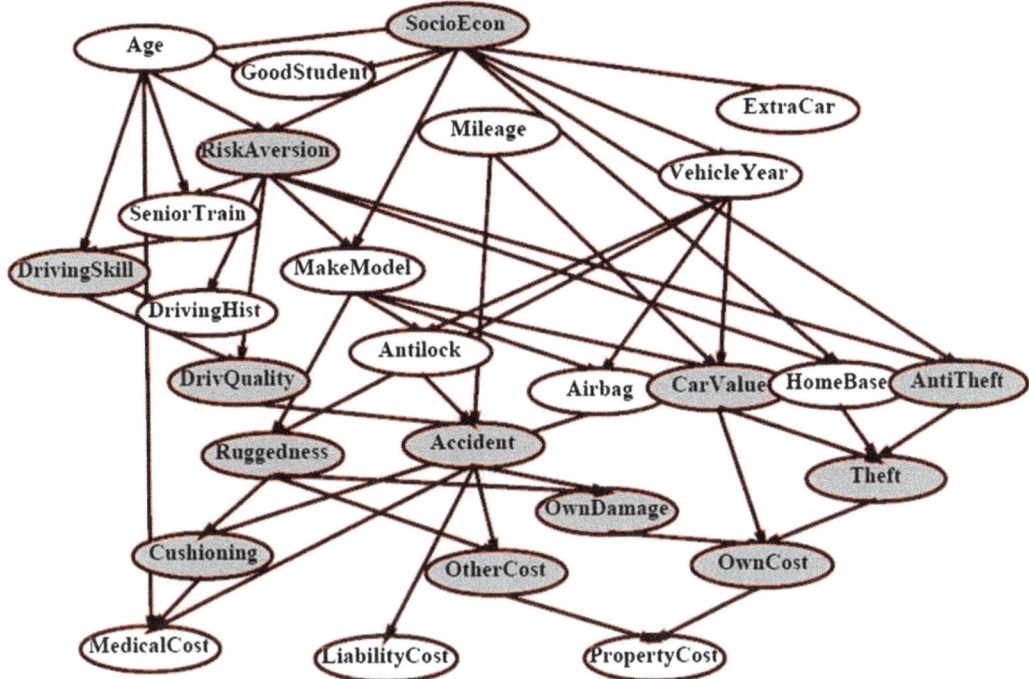

Fig. 6.6 Bayesian network for car insurance

6.5 Priors and Prejudice

To Bayesians, there is really no such thing as an "objective" finding. The action taken by the Bayesian is dependent on her own personal utility function, and the underlying probabilities are colored by her prior beliefs.

This can lead to legal and ethical problems. For example in a UK court it is illegal to make use of prior knowledge, because the legal model requires that all evidence used to make the jury's decision is presented in a public courtroom. Hence jury members are not allowed to research suspects on the internet or read about them in newspapers. Use of priors in a judicial setting is literally "pre-judice" and is the origin of this word.

Similarly, car insurers are legally prevented from using some types of prior information, such as driver gender, to predict accident rates and to set premiums. This can lead to quite strange situations where, for example, an insurer might argue they were making predictions based only the drivers' preferences for pink and blue clothing rather than their actual gender.

- Are these good laws? Why or why not?

Use of priors is also becoming a hot topic in autonomous vehicle research. In the UK Advanced Driving test, human drivers are taught to make predictions about probable behavior of other road users based on whatever information is available to them. For example, suppose that drivers of a particular brand of car have historically caused many incidents. We could say that the visual presence of this brand on a car acts as an "Antisocial, Unpleasant Driver Indicator". When we see an Antisocial, Unpleasant Driver Indicator on a vehicle ahead of us on a motorway, we then assume there is a higher than usual probability that an incident may occur, so we drive more defensively. Similar reasoning can be applied to all types of "clues" observed on the road, which might include vehicle class, size, color, registration year; and more controversially, driver and pedestrian age, gender, skin color, clothing or hair styles.

- Should we program self-driving cars to make similar discriminations?

6.6 Causality

Frequentist Statistics has historically had many problems with the concept of causality. It is common to hear that "correlation is not causation" and for frequentists to say that they only deal with correlations. In contrast, Bayesian theory – and specifically Bayesian Network theory – has developed rigorous mathematical tools to handle causality, and is happy to use them.

Like most things, causality is just a human perceptual construction. Modern theories of Physics do not refer to it as a fundamental ontological category. Since Einstein's theories of space-time, time has been viewed as just another dimension, and theories of Physics are generally "reversible", meaning they are indifferent to the direction in which time is considered to run.[5] Causation as a psychological category appears to us to be bound to the notion of the "arrow of time" in which causes precede effects – but two passengers on passing near-light-speed trains will see supposed "causes" and "effects" appear in different orders from one another. If Physics does provide any insight into a purely psychological theory of causation then this comes from its theory of Entropy, which shows that given a highly ordered starting point (such as the big bang, or the birth of a brain), time will at least *appear* to flow to observers in the direction that decreases this order. However, even if Causality is just a perceptual fiction arising from the flow of entropy in our own neurons, it still appears to be a pragmatically useful one, and building models in terms of it may enable us to make transportation function better, which is all that really matters.

"Real scientists" (as opposed here to both data scientists and fundamental physicists) have always known that they do work with causal reasoning. They usually see the purpose of their research as understanding causes and effects, especially in sufficient detail for engineers to use to prescribe casual interventions in the world to bring about desired effects. For example, suppose we are a car insurance company and we find a correlation between listening to classical music and being a good driver. Statistics alone cannot tell which causes which. But as (financial) engineers, we want to know the causation direction – should we offer a discount to drivers who agree to install a connected device that streams music of our choice into their cars? "Real scientists" know that in order to make inferences about causation from a system, it is necessary to first put some causation into the system. This is called an experiment. For example, if we cause 100 drivers to listen to Mozart and another 100 to listen to Metallica, and notice that they drive equally well, then we can conclude that the music does not cause the driving performance. Perhaps there is then some other underlying cause of both variables, such as demographics, which explains them.

Bayesian Network theory has a beautifully simple, algorithmic, understanding of this process. While it is possible to construct Bayesian Networks that are not causal, we almost always design them – intuitively – with causal semantics in mind. Precisely, this means:

---

[5]Even the collapse of quantum wave functions can be set up in this way in relativistic Quantum Field Theory, which goes beyond non-reversible Quantum Mechanics.

6.6 Causality

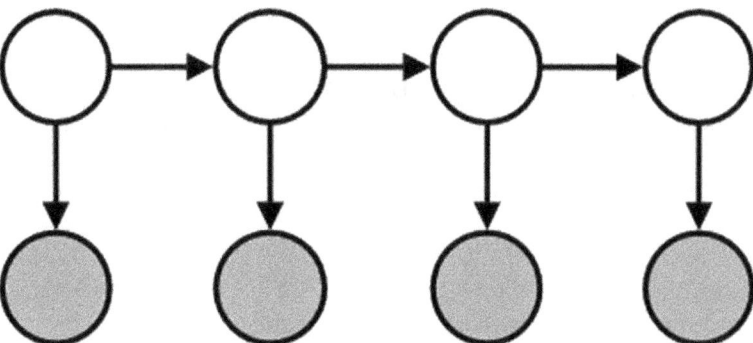

Fig. 6.7 Temporal Causal Bayesian Network. Time runs left to right. An unobserved physical process evolves over time, causing an observation at each time

- If the experimenter was to force the value of a node X – written as "do(X)":

  - then the value of node X becomes fixed to that value, irrespective of the other nodes,
  - and the links from X to its parents are cut, making them independent of X and of each other.

If we construct our networks with these semantics, then they are preserved throughout inference and it becomes possible to make inferences about causation throughout the network. For example we can calculate with terms such as $P(Y|do(X)) \neq P(Y|X)$.

Define a *proximal* cause of X as any parent of X in the causal Bayesian network. Define a *distal* cause of X as any node Y such that there exists a directed path from Y to X in the network. For example, the position and speed of a car hitting a pedestrian may be proximal causes, while the driver's aggression and lateness for work may be distal causes which act via the position and speed. Distal causes include proximal causes in this definition.

Our temporal intuition about causality – that causes precede their effects – is captured automatically by temporal Bayesian network models such as in Fig. 6.7. For example, this network models the road-user tracking problem described above if the unobserved physical process is the vehicle's state, and the observations are sensors such as CCTV images.

It is possible to infer causality from data sets if they contain some other causal information initially. Such information may be in the form of statements like "X is a proximal cause of Y", "X is a distal cause of Y", "X is a not proximal cause of Y", "X is not a distal cause of Y", or even prior probabilities about such statements. Typically such causal information is obtained when data is collected from a controlled *experiment*. If the experimenter herself causes X to take guaranteed values, then she can guarantee that no other variable is a distal cause of X. In some cases it is possible to provide causal input information from purely passive data sets by making assumptions based on knowledge of what the data represents and how it was collected. For example, passive data of driver feet positions on car pedals and CAN[6] bus logs of their speeds and acceleration could be safely assumed to have the pedals as causes of speeds and accelerations, but not the other way around.

At its simplest, causal inference can then be done by hypothesizing many different, randomly-generated Bayesian network models, and testing each one's fit against the data, including the fit of its causal input statements. Any hypothesized model whose structure violates a causal input statement can be discarded from consideration, i.e. assigned a probability of zero. If causal input statements are probabilistic, then these probabilities are multiplied into the prior probability of the model. In principle, the space of all possible models can be tested against both the data values and causal input statements

---

[6]Controller Area Network, a common networking standard for communication between electronic devices inside a vehicle.

in this way, and the model posteriors computed. In practice, the space is usually very large so heuristic search methods are used to sample from it instead.

The issue of causality is important for Data Science because it describes clearly how it differs from "real science". Remember, "real scientists collect their own data" through causal experiments, while data scientists usually do not. "Real scientists" can make causal inferences and models by using the causal input information as well as the numeric values from their experimental data. Data Science works by re-using old data from other peoples' experiments and is not usually able to do this.

In some cases, a data scientist will switch to being a "real scientist" for some, usually small, part of her work. This occurs if she is able to somehow find some causal data like a "real scientist". This could happen in several ways such as:

- Re-using old data that was itself collected in a causal way, such as controlled trials, and which comes with initial causal information attached as meta-data.
- It is sometimes possible to rule out (rather than confirm) causal hypotheses from correlation data alone.
- Sometimes the data scientist must turn into a full "real scientist" and go out and collect new real causal data! This might be a much smaller sample than the non-causal big data that she is trying to make causal inferences about. For example, she might have a terabyte of data about speeds, personalities, car radio use, and GPS traces from millions of drivers, with no idea of what causes what. She then runs a small laboratory study, causing just 100 drivers to listen to music and testing for causation of driving speed in response. The result of this may be a causal statement that music is, or is not, a distal cause of speed. This statement can then be used as a prior when fitting models to the larger set of millions of drivers, in a hybrid of "real" and "data" science.

In short: you *can* read causal knowledge out from inference if you put causal knowledge into it.

- Figures 6.8 and 6.9 show two purported causal theories of climate change. Do you find either of them convincing from the data alone? If your logic differs between the two cases, explain exactly why, using causal model concepts.

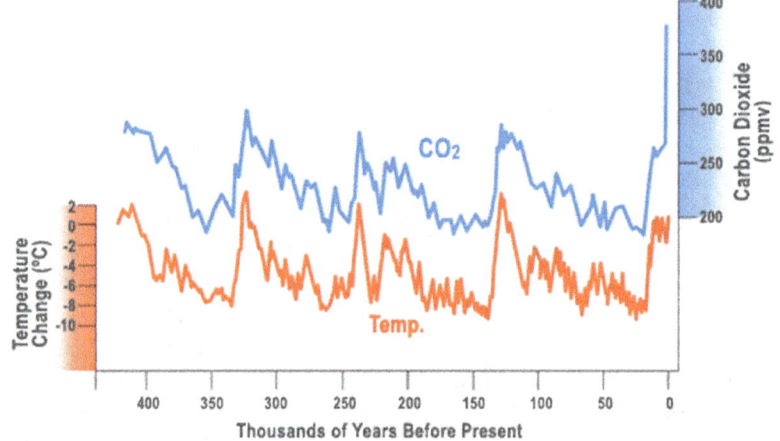

Fig. 6.8 A causal explanation for climate change?

Fig. 6.9 A causal explanation for climate change?

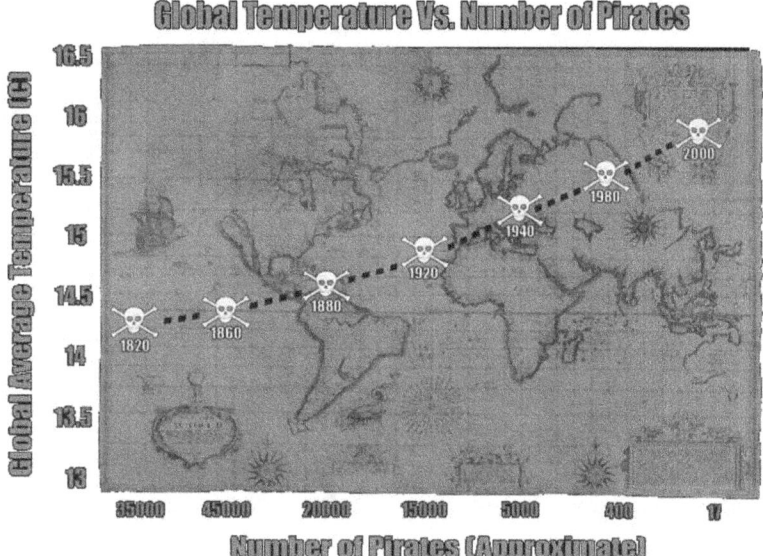

6.7 Model Comparison and Combination

What happens if we can think of two or more rival models for some data? For example, we might propose a new model of the accident data that allows for three different environment states rather than two.

Bayesian theory contains a very beautiful way to compare models against each other. Frequentist statistics often has difficulty with this concept, for example often proposing two models where one is a "null hypothesis" that is designed to be obviously wrong or uncompetitive with the proposed model. This is based on a mis-understanding of the notion of "hypothesis rejection" from Karl Popper's philosophy of science. What Popper actually said is that science should be a competitive process, with many models fighting against one another to best explain the data. Over time new models are proposed and replace the old ones because they are better. Null-hypothesis advocates mis-read this as meaning that they should propose two of their own models, and show one to be false, in order to accept the other. Bayesians read Popper correctly. When Bayesians compare models, it is a real fight between serious contenders. There are no null-hypotheses. Rather, multiple good models are proposed either by the same author, or more interestingly, by rival authors, and are tested head-to-head against each other to see which is best. This is, in theory, easily done for models M and data D via,

$$P(M|D) = \frac{P(D|M)P(M)}{P(D)}.$$

In practice this is hard to compute ($\mathcal{NP}$-hard) but again is approximated by algorithms and computer software such as PyMC3. As a user you can just ask for the result without worrying how it works.

One of the strangest aspects of Bayes is that the concept of "the one true model" tends to disappear along with the "true parameter values". You can usually make better action selections by averaging[7]

---

[7] Usually you can do even better than this "Bayesian Model Averaging" using other types of model combination. Averaging is only optimal if you are sure that your hypothesis set contains the ground truth rather than just approximations (P. Domingos, *Bayesian Averaging of Classifiers and the Overfitting Problem*, International Conference on Machine Learning (ICML), 2000).

over a whole set of models,

$$\hat{a} = \arg_a \max \sum_M \sum_s U(a, s, M) P(s|D, M) P(M|D),$$

even including ones which are not very good and would have been "falsified" under Popper's philosophy of science. Bayesianism is a pragmatist philosophy – there is no "truth", other than utility!

6.8 Exercises

PyMC3 performs MCMC inference in Bayesian networks.[8]

6.8.1 Inferring Traffic Lights with PyMC3

Here we compute the traffic light example above using PyMC3.

```
import pymc3 as pm
from scipy import stats
import matplotlib.pyplot as plt
with pm.Model() as trafficLight_model:
    # define our data
    data = stats.bernoulli(0.2).rvs(100000)
    theta = pm.Beta("theta", alpha=1.0 , beta=1.0)
    color = pm.Bernoulli("color",p=theta,observed=data)
    start = pm.find_MAP()      #start MCMC from MAP
    step  = pm.Metropolis()    #MCMC
    trace = pm.sample(1e4, start=start, \
            step = step, model=trafficLight_model)
    # plot results
    pm.traceplot(trace[0:]);
    plt.show()
```

6.8.2 Inferring Accident Road State Change with PyMC3

Here we compute the accidents example above using PyMC3. This is a more complex model than the traffic light because it features a "custom" node, *rate*, whose switching behaviour is not modelled by any standard PyMC3 node type, and has to be programmed explicitly. (The lines invoking a library underlying PyMC3 called Theano are effectively "magic" and don't need to be understood by most users. This example can easily be modified to create other customer nodes by keeping the Theano lines in place, and modifying only the inside of the *rate* function, and is provided for this reason.)

```
import pymc3 as pm
import numpy as np
import theano.tensor as t
```

---

[8]Thanks to ITS Leeds student Panagiotis Spyridakos for porting this code to PyMC3.

```
import theano from theano.printing
import pydotprint
import matplotlib.pyplot as plt
with pm.Model() as inferAccidents_Model:
    data=np.array([ 4, 5, 4, \ 0, 1])   #paste data here
    switchpoint = pm.DiscreteUniform(
        'switchpoint', lower=0, upper=110)
    early_mean = pm.Exponential('early_mean', 1.0)
    late_mean = pm.Exponential('late_mean', 1.0)
    #defining a custom node
    @theano.compile.ops.as_op(itypes=[t.lscalar, \
        t.dscalar,t.dscalar],otypes=[t.dvector])
    def rate(switchpoint, early_mean, late_mean):
        out=np.empty(len(data))
        out[:switchpoint] = early_mean
        out[switchpoint:] = late_mean
        return out.flatten()
    accidents = pm.Poisson('accidents', \
    mu=rate(switchpoint, early_mean, late_mean), observed=data)
    pydotprint(inferAccidents_Model.logpt)
    start = pm.find_MAP()
    step = pm.Metropolis()
    trace=pm.sample(1e4, start=start, \
    step = step, model=inferAccidents_Model)
pm.traceplot(trace[0:]);
plt.show()
```

The result of this will be histograms showing posterior beliefs about the switching time, and the Poisson parameters of the two environment states.

6.8.3 Switching Poisson Journey Times

Use the accidents model above, and data from Derbyshire County Council, to make a switching Poisson model of the number of journeys on a sample origin-destination route, switching on time of day (5 min windows). Experiment with variations on this model, such as using more that two environment states.

Design on paper only or in PyMC3, a Bayesian network model for another DCC application.

6.9 Further Reading

- Stone JV (2013) Bayes' rule: a tutorial introduction to Bayesian analysis. Sebtel Press (Easy. The best place to start learning Bayes)
- Bernardo, JM, Smith AFM (2001) Bayesian theory. vol 221 (Advanced. The brilliant, definitive book on Bayes.)
- Pearl J (1988) Probabilistic reasoning in intelligent systems: networks of plausible inference. Morgan Kaufmann (The brilliant, definitive book on Bayesian Networks)
- Pearl J (2009) Causality. Cambridge University Press (Definitive book on causality)

- Glymore G, Cooper G (1999) Computation, causation, and discovery. AAAI Press (Includes debates about whether causal inference works)
- http://docs.pymc.io/index.html (Full PyMC3 reference)
- R. vs Clerk. 2003. EWCA Crim 1020. Case No: 200203824 Y3

Machine Learning

"Machine learning" in its current popular use refers to models which learn to make classifications directly as functions of the data, without using any generative or causal models. We are given some known data pairs $\{(\mathbf{x}_n, c_n)\}_n$, where $\mathbf{x}_n$ are features and c_n are discrete classes, then are asked to find the class c_m of single new data point $\mathbf{x}_m$ through some function $c_m = f(\mathbf{x}_m, \theta)$. Machine learning consists of finding the parameters θ.

This is more precisely known as "Discriminative Classification",[1] and is also known informally as "black box" modeling, which emphasizes its aim to give useful outputs without using anything understandable or meaningful on the inside.

Historically, classification and AI research more generally have oscillated between "neat" theory-driven, and "scruffy" black-box approaches, typically around once per decade. Generative inference is a "neat" method, dominant in the 2000s, while the last few years have seen the latest return to discriminative methods via hierarchical linear-in-parameters regression (also known and hyped in popular literature as "neural networks" or "deep learning"). All these methods have existed for over fifty years (Duda and Hart, 1973) but they go in and out as fashion largely depending on the state and price of practical computing hardware. The current trend has been enabled by price falls in parallel GPU hardware, especially from the company NVidia which has re-purposed its cheap, consumer video game products to run these models very fast.

7.1 Generative *Versus* Discriminative Vehicle Emissions

The previous chapter on Bayesian Inference made use of generative models to understand observed data. There, the models described the causal processes which give rise to the data. These processes were parameterized by hidden variables, such as the θ of the traffic light. Bayesian theory took the generative model, $P(D|\theta)$ and used Bayes' rule to invert it to compute $P(\theta|D)$.

We may use generative models for classification too. Suppose that we have emissions level data, $\mathbf{x}$, from two classes c of vehicle, petrol (p) and diesel (d) cars, and would like to recognize future vehicle

---

[1] In academia, "Machine Learning" has been used to name various areas of related research by a loose community whose priorities and popularities change over time. Discriminative Classification is one such topic, which has crossed over from this research community into practice, and taken the larger community's name with it. Academic "Machine Learning" has also explored non-probabilistic symbolic category learning, temporal reinforcement learning, game theory, generative Bayesian inference and other topics throughout its history.

types from their emissions detection signatures. Assume that the emissions data are two-dimensions, having NOx and CO2 components,

$$\mathbf{x} = [NOx, CO2].$$

Using generative Bayesian modeling, we might model each type of vehicle as having its own two-dimensional Gaussian distribution over emissions level,

$$P(\mathbf{x}|p) = N(\mu_\mathbf{p}, \Sigma_\mathbf{p})$$

$$P(\mathbf{x}|d) = N(\mu_\mathbf{d}, \Sigma_\mathbf{d}),$$

where μ and Σ are mean and covariance matrices, and,

$$N(\mu, \Sigma) = \frac{1}{(2\pi)^{dims/2}|\Sigma|} \exp\left(-\frac{1}{2}(\mathbf{x}-\mu)\Sigma^{-1}(\mathbf{x}-\mu)\right).$$

If we assume that the covariance matrices are equal, and circularly symmetric,

$$\Sigma_p = \Sigma_d = \begin{bmatrix} \sigma & 0 \\ 0 & \sigma \end{bmatrix},$$

then these Gaussians will appear as circular densities in the 2D (NOx, CO2) state space. Figure 7.1 shows these, together with 100 samples drawn from each Gaussian, and a separating line between the classes.

To fit the two models to data, we have to fit five parameters: two mean vector components for each class and a single standard deviation, σ. To classify the type of a new vehicle from its observed emissions, we would them compute $P(c = d|\mathbf{x})$, $P(c = p|\mathbf{x})$ under these parameters, via Bayes theorem, and choose the highest valued one as the classification to report.

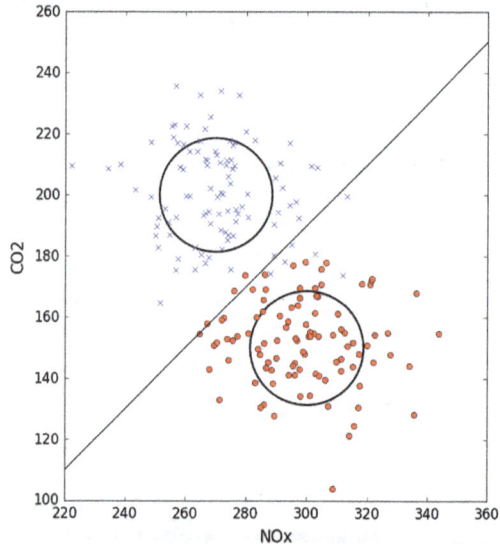

Fig. 7.1 Emissions discrimination

However, for this problem there is a much simpler and faster way to obtain the same classifications. Consider the 2D state space above. Which regions of the space will be classified as petrol and which as diesel? In other words, in which regions does which Gaussian have the higher value? Because we assumed the Gaussians to be circular, they effectively partition the space into two classification regions, whose border is a straight line as shown above. We know that straight lines can be parameterized by just two variables – an intersection and a slope. So instead of fitting five parameters to the data, we could get by with fitting just two, to describe the same classification boundary. Computing which side of a straight line the new point lies on is much quicker than running inference in the generative model, for example it could run in real-time where PyMC3 cannot. This is the idea of discriminative models: rather than model the generative parameters, they model only the shape of the classification boundary.[2]

7.2 Simple Classifiers

7.2.1 Linear Discriminant Analysis (LDA)

It can be shown (see Duda and Hart) that the optimal separating boundary for the two-class Gaussian problem is linear as in the figure above. Therefore, we can find the optimal separating line via the optimization problem,

$$(\mathbf{w}, t) = \arg_{\mathbf{w},c} \max \sum_n ((\mathbf{w}.\mathbf{x_n} > t) = c_n),$$

where $c_n \in \{0, 1\}$ are the true classes of data $\mathbf{x_n}$ represented in binary ($0 =$ petrol, $1 =$ diesel), and the Boolean value of the equality is also interpreted as integers ($0 =$ False, $1 =$ True). This is a linear optimization problem which has an analytic solution. You don't need to know the derivation of this because it is implemented in many software packages for you.

- When will this be likely to work? When will it fail?

7.2.2 Nearest Neighbor

Often your data is not linearly separable and will require something more powerful to classify. The simplest classifier is often to use nearest neighbors. Here, when a new data point arrives, x, we simply measure its distance from *every* known classified point in the database. We find the single closet match in the database, read its class, and assign the same class to the new point,

$$c = c_m,$$

where

$$m = \arg_n \min |\mathbf{x} - \mathbf{x_n}|^2.$$

---

[2] An interesting and – as far as the author knows – still open research question asks in what cases can particular types of generative models be "compiled" into simple discriminative classifiers in this way. For example, given an arbitrary Bayesian Network, is there an algorithm which computes its decision boundary as a fast, analytic equation which can then be used as a classifier? More heuristically, the "Helmholtz machine" described the idea of using samples drawn from a generative model to train an arbitrary flexible classifier such as a hierachical sigmoid regression model.

In this case we have used an L2 norm (i.e. the Pythagorean distance between the two vectors) as the distance measure, but other distance measures could also be chosen. (This is a simple case of a large class of methods known as *kernel*methods, which compare the new point to some or all of the old data directly rather than via any compressing functions of the data such as separating plane parameters.)

- When will nearest neighbor be likely to work? When will it fail?

7.2.3 Template Matching

If you don't have the computation time available to use nearest neighbor, a closely related method is to compute a single "prototype" or "template" model of each class. This can be done, for example, simply by taking the mean of each class,

$$\bar{x}^{(c)} = \frac{1}{\sum(c_n = c)} \sum_n \mathbf{x}_n (c_n = c),$$

then classifying by,

$$c = \arg_c \min |\mathbf{x} - \bar{\mathbf{x}}^{(c)}|^2.$$

This method is commonly used for Automated Number Plate Recognition, for example underlying the M25 motorway study of Chap. 1. In ANPR, a separate process first extracts the licence plates, then normalizes them to a standard pixel size. If we know that there is a standard size string of characters on the plate, we can then divide this image into per-character sections, as show by the black box around the "S" in Fig. 7.2.

In this example, each character is now contained in a box of 20 × 30 pixels. These 600 pixels can be quantized to (0 = white, 1 = black) values, and inserted into a 600 element vector. Such vectors are used during training to build the mean templates, and at runtime to compare against the templates.

- When will nearest neighbor be likely to work? When will it fail?

7.2.4 Naïve Bayes Classification

Naïve Bayes is a discriminative classifier which derives from a simplified generative Bayesian model of the data. It assumes first that the observed features are independent of each other given the class,

$$P(\mathbf{x}|c) = \prod_i P(x^{(i)}|c),$$

and second that all classes have equal priors,

$$P(c_i) = \pi, \forall i.$$

7.2 Simple Classifiers

Fig. 7.2 ANPR machine vision stages

Under these assumptions, we have (via Bayes theorem),

$$P(c|\mathbf{x}) \propto \prod_i P(x^i|c),$$

where the $P(x^i|c)$ can be fit from historical data individually.

This classifier can still be used even when our true generative model includes dependencies between the features. The independence assumption is used just to build a simplified discriminative classifier rather than a full generative model.

- When will this be likely to work? When will it fail?

7.2.5 Decision Trees

Decision trees are simple classifiers which operate similarly to the game of "twenty questions". In this game, someone picks a famous person and you are allowed to ask up to 20 questions to find out who the person is. For example,

1. Is the person alive? No.
2. Did they live after the year 1500? Yes.
3. Did they live before 1900? Yes.
4. Were they male? Yes.
5. Was he an artist? No.

6. Did he work with transport? Yes.
7. Did he design vehicles? Yes.
8. Was he Robert Stephenson? No.
9. Did he work with mass production? Yes.
10. Is he Henry Ford? Yes.

Once you learn to play, you will choose questions that are maximally informative at each step. In the case of yes/no questions, this means that you should be a priori 50% certain that the answer will be "yes" or "no". Decision trees will do exactly this in the case of Boolean data. More generally, if they are used with continuous valued data, they will find maximally informative questions like,

1. is $x<3$? Yes
2. is $y<0.5$? No.
3. is $x<0.2$? Yes.
4. is $y>0.8$? Yes
5. I think the class is *petrol car*.

as shown in Fig. 7.3 from a computational output. Formally, "maximally informative" means that according to the historical data, obtaining the answer to the questions will reduce the entropy of our posterior belief about the class. At each step, most decision tree algorithms choose the single question to maximally reduce entropy. This is known as a "greedy algorithm" as it does not take account of relationships between questions in a sequence, but is usually good enough.

- When will this be likely to work? When will it fail?

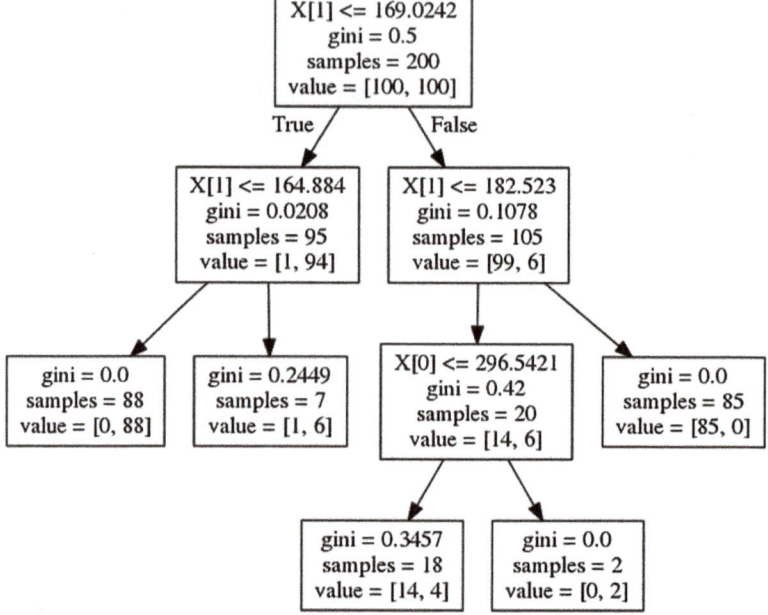

Fig. 7.3 Decision tree

7.3 Neural Networks and "Deep Learning"

All of the above generative and discriminative methods can be thought of as partitioning the feature space into discrete partitions with class labels. They all impose different constraints and assumptions on the shape of the partitioning boundary. These are effectively priors on the boundary shape. It is natural to ask if it is possible to build discriminative classifiers having many parameters, θ, which allow for completely arbitrary boundary shapes to be described. For example, suppose we have the 3-dimensional feature sets for two classes shown in Fig. 7.4. We can see that there does exist some very curved, irregular, boundary plane that could be used to make almost perfect discrimination.

There are, in fact, many classes of parametric functions that are able to model arbitrary boundaries as,

$$c = F(\mathbf{x}; \theta).$$

Any particular function from one of these classes will necessarily have a finite number of parameters, and therefore have some limit on what it can model. But considered as classes of functions, with members having more and more parameters, it can be proved that as the number of parameters increases, then any arbitrary boundary can be modelled. Locally optimal values of the parameters can be found via gradient descent steps (of size α),

$$\Delta\theta = -\alpha \nabla_\theta E,$$

where E is a measure of the total error between the model's prediction of the training data's classes and its true classes, such as,

$$E = \sum_n (c_n - F(\mathbf{x}_n; \theta))^2.$$

While there are a great many such function classes, one class which is used traditionally and which has nice computational properties is the class of hierarchical linear-in-parameters regression (HLIPR) functions. (These are sometimes referred to, especially by the media and by people trying to sell them, as "neural networks", "deep learning", or "deep neural networks", "artificial intelligence", "electronic brains" etc.)

HLIPR is based on linear-in-parameters regression, which finds weights w_i to approximate a continuous-values function $y(\mathbf{x})$,

Fig. 7.4 Nonlinear decision boundary

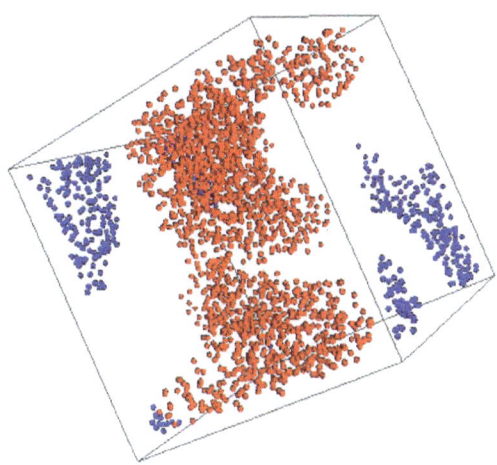

Fig. 7.5 Neural network structure

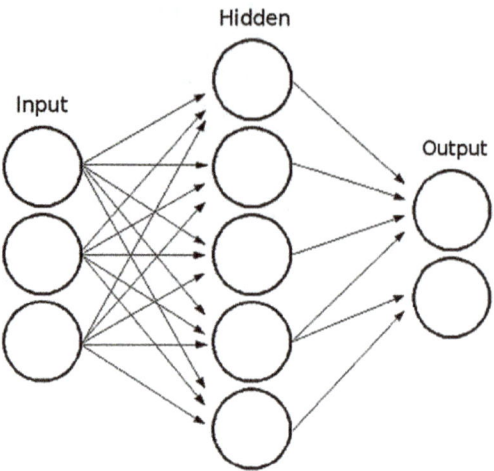

$$y(\mathbf{x}) \approx \hat{y}(\mathbf{x}) = f\left(\sum_i w_i x_i\right),$$

where x_0 is often fixed as $x_0 = 1$ (known as a "bias" or "affinating" term) and inputs represented in the $x_{i>0}$; and f is some non-linear function, such as the commonly used sigmoid,

$$f(v) = \sigma(v) = \frac{1}{1 + e^{-v}},$$

or the rectified linear unit (ReLU) function,

$$f(v) = ReLU(v) = max(0, v).$$

HLIPR uses many of these functions, arranged into layers, with the output of one layer feeding into the next, which may be viewed as nodes in a graph, as in Fig. 7.5.

This is not a Bayesian network – the arrows point the opposite way, from effects to causes, because we are doing classification rather than generative modelling.

The name "linear-in-parameters" means that f is a non-linear function of a 1-dimensional real input, but that input is itself a linear weighted function of multiple inputs. If we made such a network using only a linear function, rather than one which is linear-in-parameters, then the multiple layers would be redundant and we could write the whole model simple as one linear regression. It is the use of the non-linearity in the nodes that allows complexity to be introduced into the model. HLIPR is a infinite class of functions, each comprised of layers of nodes, with different numbers of layers and nodes per layer, and with different choices of linear-in-parameters functions. It can be shown that this class is able to model any (non-pathological[3]) function, with some finite number of layers and units per layer.

For the final layer, we treat the Boolean valued $c \in 0, 1$ as a real, $y \in [0, 1]$.[4]

The goal is thus to model some function $y(x) \in [0, 1]$ from examples.

---

[3] Pure mathematicians might enjoy finding strange, non-continuous, functions to break this; data scientists don't care.

[4] HLIPR can more generally be used to approximate any real-valued function rather than just binary classification, hence the "regression" in its name rather than just "classifier".

7.3 Neural Networks and "Deep Learning"

We will use a notation convention: let jth layer have J nodes indexed by j, with layers connected as $k \to j \to i$. For each layer's outputs,

$$y_j = f(x_j) \tag{7.1}$$

$$x_j = \sum w_{kj} y_k, \tag{7.2}$$

where f is any differentiable function of one variable y_j is always a linear-in-the-parameters function. For an arbitrary error function E we wish to set weights to locally minimise it by moving in directions,

$$\Delta w_{jk} = \alpha \frac{\delta E}{\delta w_{kj}}. \tag{7.3}$$

Define,

$$\Delta_j = \frac{\delta E}{\delta y_j} \frac{\delta y_j}{\delta x_j}. \tag{7.4}$$

Then by chain rules,

$$\frac{\delta E}{\delta w_{kj}} = \left[\frac{\delta E}{\delta y_j} \frac{\delta y_j}{\delta x_j} \right] \frac{\delta x_j}{\delta w_{kj}} \tag{7.5}$$

$$\frac{\delta E}{\delta w_{kj}} = \Delta_j y_k. \tag{7.6}$$

$\frac{\delta y_j}{\delta x_j}$ is the simple gradient of the activation function,

$$\frac{\delta E}{\delta y_j} = \sum_i \frac{\delta E}{\delta y_i} \frac{\delta y_i}{\delta x_i} \frac{\delta x_i}{\delta y_j} \tag{7.7}$$

$$\frac{\delta E}{\delta y_j} = \sum_i \Delta_i w_{ji}. \tag{7.8}$$

Plugging this into the definition of Δ_j gives the 'backpropagation' iteration,

$$\Delta_j = \frac{\delta y_j}{\delta x_j} \sum_i \Delta_i w_{ji}. \tag{7.9}$$

The final layer is a special case, because we know what the desired output label c is from the training data. E is some function only of this top ith layer over training data indices d, such as,[5]

$$E = \sum_d \sum_i \frac{1}{2} (c_i^{(d)} - y_i^{(d)})^2. \tag{7.10}$$

---

[5] As we are doing classification rather than regression, $y_i = y$, $c_i = c$ and $i = \{0\}$, i.e. there is only one output node. For more general regression, the output may be high dimensioned with i equal to a larger index set. The generalized regression version is given here in case you ever need it.

This least squares error function is theoretically motivated by a Gaussian noise assumption, but is really chosen to make the maths nicer because,

$$\frac{\delta E}{\delta y_j} = \sum_d \sum_i (y_i^{(d)} - t_i^{(d)} y_i^{(d)}). \tag{7.11}$$

If we choose the particular sigmoid function for the nonlinearity,

$$f(x) = \sigma(x) = \frac{1}{1+e^{-x}} \tag{7.12}$$

$$\sigma'(x) = \sigma(x)(1 - \sigma(x)) \tag{7.13}$$

$$\Delta_j = \sigma(x)(1 - \sigma(x)) \sum_i \Delta_i w_{ji} \tag{7.14}$$

$$\Delta w_{jk} = \alpha \Delta_j y_k. \tag{7.15}$$

This simple update is what actually gets programmed in most neural network software, together with the back-propagation update. Together, these are recursive computations which compute back-propagation terms Δ_j for each layer, beginning at the output layer and working backwards towards the input layer (hence "back-propagation"). This pattern of computation is the opposite of the forward computation of the network, which begins with the input and updates layers moving towards the output. Also at each layer during back-propagation, the parameters w_{jk} in that layer get updated using the Δ_j. This nice computation structure emerges because of the use linear-in-parameters functions. If an arbitrary non-linear f was used then we would not get the back-propagation structure. (Proof: exercise).

7.3.1 Parallel Computing Back-Propagation

The back-propagation idea is very old, dating from at least the 1970s and arguably to the 1960s or even to Alan Turing's 1948 work on "B-type machines". The basic idea of adjusting a parametric model with gradient descent is obvious, but the use of linear-in-the-parameters functions to make the computation fast and simple is the clever discovery. However even with the back-propagation speedup, serial computers from the 1960s up to the present decade have still only had the raw processing power to optimize parameters for models having a few (e.g. three) layers and a few (e.g. a hundred) units per layer. These have been suitable for proving the concept and solving simple classification problems. These models are just a small part of the infinite class of hierarchical linear-in-parameters regression functions.

During this decade, implementation of back-propagation has become possible for much larger members from the class, for example having 10 layers with thousands of units per layer. This has been due to:

- Mass production of consumer graphics units for gamers (thanks NVidia!)
- Cheap availability of networking and inter-networking to enable parallel PC computation (thanks Cisco!)
- Falling prices of computing hardware in general (thanks China!)

7.3 Neural Networks and "Deep Learning"

The back-propagation equations' structure, based on nodes in a graph, lends itself very nicely to parallel computation implementation, where nodes can live on different computers or processing components, and send Δ_j terms as messages over networks or buses between one another. Software implementations of back-propagation are widely available to take advantage of this structure using the above hardware technologies. Some popular packages include Caffe (running on CUDA and soon on OpenCL), Google TensorFlow (running on the same); and DL4J (DeepLearning4Java, running on Hadoop clusters). (OpenCL is a programming layer for many GPU graphics cards and also for FPGA and other parallel computers; CUDA is similar but is only for NVidia's propitiatory graphics cards.)

Deep learning is primarily a contemporary marketing term for HLIPR. If it has any technical meaning then it would be best used to refer specifically to the use of parallel computation to compute HLIPRs (which typically enables larger network with more layers to be trained). This is a similar form of definition to "big data", which we will also define in terms of parallel computation rather than size.

7.4 Limitations and Extensions

Boundary Assumptions. The simple classifiers all make different, particular, assumptions about the shape of the classification boundary. For LDA, it is linear. For Decision Trees, it is "boxy". For templates and nearest neighbor, it is "blobby" and centered around exemplars or means. All classifiers assume the space is "smooth", in that a point which is "near" another point will be classified the same.

Defining Similarity. This leads to the question of what is "nearness" or "similarity" of two entities? All these classification models assume that entities are represented by vectors in well-defined, finite-dimensional, feature spaces, which come with distance metrics. It is not always the case that this assumption holds. As humans, we are able to perceive many pairs of entities as "similar" when they are not presented to us as vectors in feature spaces. For example, is a Porsche Boxter more "similar" to an MG TF or to an Audi TT? In what sense? Perhaps our human notion of similarity gives greater consideration to "features" such as engine size or acceleration; or to social factors such as being driven by idiots or posers? Some people (including some of their drivers) may not perceive some of these features at all, or will choose to invent and perceive other features that are important to them. There is an infinite number of possible features that could be constructed and weighted in similarity judgments, and feature-space classifiers to not give any consideration to this – they assume that the choice has already been made by an intelligent human.

Scene Analysis. As with generative models, we don't always know what a single entity is in the first place, which is needed to begin any classification process. For example, if we wish to classify pedestrians in a self-driving car's camera images, we need some other method to find where the pedestrians are before we can calculate features to describe and classify them. This is a chicken-and-egg problem – how can we build a classifier to discriminate between pedestrians and non-pedestrians if we don't know where to point it at in the first place? In practice this is usually solved by "priming" hypotheses. This uses some simple, fast, feature of the scene to give a first rough guess about pedestrian-ness, which is applied to every possible location in the scene. Then the interesting candidates are passed to the more computationally intensive classifier for more detailed analysis. For visual tasks, it has been found that "convolution neural network" topologies and functions tend to work well, which restrict the data available to each function to local visual windows.

Over-Fitting. HLIPR models can have arbitrarily large numbers of parameters, for example current hardware can run with millions or tens of millions of parameters. As with generative models, large parameter spaces require more data to inform them. Without enough data, optimization will over-fit the training data, and fail to classify new examples. To some extent, modern data sets are now large enough to avoid this problem. If you are working with search-engine sized data then there will be enough. It is still not well understood[6] how to theoretically determine how much data is required to inform a particular model. In practice, this question can be avoided by making use of separate training and validation data sets. The training data is used to train the classifiers, then they are tested on the validation data. If performance on the validation data is poor this suggests over-fitting has occurred, then you go back and either train with extra data or with fewer parameters to avoid it. A useful trick to discourage over-fitting in the first place is to deliberately add noise to multiple copies of the training data. Some HLIPR models do this by randomizing outputs of high-level nodes as well as the data itself, which is known as "dropout" training.

Ensembles. As with generative Bayesian Network models, it is possible to compare or combine results from multiple classifiers to get better results. This can be more important than when working with generative models, because typically we have less idea about the structure of the data or model when we use a "black box" classifier than a theory-driven generative model. A simple way to do this is just to train lots of classifiers and test them on a test set to see which is the best. There are then various ways to combine their classifications, including simple averaging or voting, Bayesian Averaging, or more complex schemes such as "bagging" and "boosting". These can be applied to multiple copies of the same simple classifier, trained on different subsets of the training data, to form complex and better-performing "ensemble" classifiers, which behave similarly to HLIPR models. A particular popular version for combining many decision trees is called "random forests". Such ensemble classifiers usually win machine learning competitions such as the NetFlix challenge and Kaggle contests.

Dredging. To "neat" Bayesians, discriminative methods can look and feel rather dirty and unpleasant. The arbitrariness of the functions used suggests worries about what the model priors should be, and the large parameter spaces suggest worries about over-fitting. The prior probability of any particular model and parameter combination being the correct one seems vanishingly small, suggesting that only very strong fits could overcome such weak priors. (How can models even have priors when they are not based on prior generative theory?) This is sometimes the case, and many studies have been accused of "data dredging" or "p-hacking"[7] where they effectively try out millions of possible random models until they find and publish one that appears to fit some data. Such models are unlikely to work on any future data! This problem can be fixed, again, by using separate training, validation, and test data sets. Any model that over-fits to the training data will simply fail on the validation data. Separate test sets should be used for final accuracy reports, and only ever consulted once at the end of the study. "Fitting to the test set" in any way is one of the worst crimes that a data scientist can be accused of, and can lead to fake publications and false models.[8] Don't ever do it!

Discriminative models can be "discriminative" in the same sense as political discrimination! Unlike generative approaches, they do not model what is actually happening in the real world, they simply look for short cuts mapping from surface features directly to probabilistic guesses about the underlying causes. A discriminative model might learn that people with brown clothes or eyes or skin are more or less likely to behave in certain ways, but this does not mean that a particular individual with this feature is going to do so, or should be treated in certain ways.

---

[6] Perhaps there is or should be some theory showing how the entropy of our belief in the best parameter values is affected by the data, and working from this to count how much information is needed to obtain sufficiently tight belief?

[7] The name derives from p-values in Frequentist statistics. Bayesians do not use these statistics but have inherited the name for the crime even when committed via Bayesian means.

[8] And for financial data, to losing a lot of other peoples' money.

7.4 Limitations and Extensions 105

Unsupervised methods can be used, as in generative models, to cluster unclassified data into classes and learn classifier parameters of those classes at the same time. For discriminative classifiers this is usually done with the EM algorithm. For example, when EM is applied to the template classifier, it infers the stereotyped cluster centers at the same time as classifying the training data into those classes, and is known as the "k-means" algorithm.

7.5 Exercises

Here is a way to generate similar random Gaussian distributed data to the NOx/CO2 example above, in Python,

```
import numpy as np
from matplotlib.pyplot import *
mu_p = np.array([270, 200])
mu_d = np.array([300, 150])
sigma = 250
Sigma = np.matrix([[sigma, 0], [0, sigma]])
xs_p = np.random.multivariate_normal(mu_p, Sigma, 100)
xs_d = np.random.multivariate_normal(mu_d, Sigma, 100)
clf()
plot( xs_p[:,0] , xs_p[:,1] , 'bx' )
plot( xs_d[:,0] , xs_d[:,1] , 'ro' )
xlabel("NOx") ylabel("CO2")
```

Use this data to fit the simple classifier models and classify new data points. Make a figure showing the classification boundaries, by classifying a whole grid of points covering the space.

For example, to perform LDA in Python:

```
import sklearn.discriminant_analysis
x = np.array([[-1, -1], [-2, -1], [-3, -2], [1, 1], [2, 1]])
c = np.array([1, 1, 1, 2, 2])
lda = sklearn.discriminant_analysis.LinearDiscriminantAnalysis()
lda.fit(x, c)
print(lda.predict([[-0.8, -1]]))
```

Here is how to build a decision tree (using an algorithm called CART) in Python, for the emissions example above. To prepare the data, we need all the input cases as a single **x** vector, and all the corresponding classes as a c vector,

```
xs = np.vstack((xs_p, xs_d))
cs = np.hstack (( zeros(xs_p.shape[0]) , ones(xs_d.shape[0]) ))
```

To make and fit a decision tree,

```
from sklearn.tree import DecisionTreeClassifier, export_graphviz
dt = DecisionTreeClassifier(min_samples_split=20, random_state=99)
dt.fit(xs, cs)
```

To draw the tree,

```
import subprocess
export_graphviz(dt, "foo.dot", ["f","g"])
subprocess.call("dot -Tpng foo.dot -o dt.png", shell=True)
```

Here we use the same data to train and use a neural network for classification,

```
from sklearn.neural_network import MLPClassifier
clf = MLPClassifier(solver='lbfgs', alpha=1e-5, \
    hidden_layer_sizes=(10, 10), random_state=1)
clf.fit(xs, cs) cs_hat_nn = clf.predict(xs)
print(cs) #show ground truth classes
print(cs_hat_nn) #show predicted classes
```

(Advanced) Find and download some ANPR character data; build and test a template classifier on it; build and test other classifiers on it.

(Advanced) Installing "deep learning" neural network software onto GPU hardware is quite complex and beyond the scope of this book. For beginners, the best approach at the time of writing is find a pre-installed setup to use and work through its tutorial documents. Some universities have such setups which they make available to external users, alternatively internet "cloud" services such as Amazon EC2 make them available for a rental fee (*https://aws.amazon.com/amazon-ai/amis/*). Find a commercial server with the library Keras or Caffe installed and use it to train larger networks. It is possible that one day soon the above *sklearn*-based code with "just work" on your GPU machine, switching to "deep learning" mode automatically, either from future *sklearn* versions or via a related project (such as *https://github.com/tensorflow/skflow*).

Think about what interesting questions could be formulated with discriminative classification on the Derbyshire data. Design and build a system. (Examples: we could create features for routes by binning numbers of detections into time bins during the day, then use an unsupervised method to cluster them into types of road such as morning and evening commute routes. Or, flows from a set of roads could be used to try to predict flows on another road.)

7.6 Further Reading

- Duda R, Hart P (1973) Pattern classification and scene analysis. Wiley. (There is also an updated 2001 edition – can you spot how much has actually changed from 28 years of machine learning research?)
- Bishop C (1995) Neural networks for pattern recognition, Oxford
- Harley A Demystifying convolutional neural networks. www.scs.ryerson.ca/~aharley/neural-networks (if you want to see what the convolutional neural net functions look like for machine vision tasks.)
- Breiman L, Friedman J, Olshen R, Stone V C (1984) Classification and regression trees, Wadsworth, Belmont, CA
- French B (1995) The subltey of sameness. MIT Press. (An argument against feature-space classification as "intelligence".)

Spatial Analysis

8

> *"Everything is related to everything else, but near things are more related than distant things."*
> — Tobler's First Law of Geography

Spatial data is generally characterized by Tobler's First Law of Geography. The Law says that geographic data is spatially *smooth,* and that we can obtain information about an unobserved point (x, y) from observations of nearby points, $(x + \Delta x, y + \Delta y)$. This is quite often the case in non-spatial tasks too, for example the whole theory of Discriminative Classification is based on the assumption that entities which are close in feature space are "similar" and likely to be of the same class. We also saw gradient descent algorithms for optimizing classifier parameters, including back-propagation, which make a similar assumption about the smoothness of error functions in parameter space. For those methods, it is somewhat debatable whether and why such assumptions always hold. But for geographical data we can usually be much more confident. This is because most geographic causal processes – including both physical and human processes – have some spatial location of their cause, and their effects tend to dissipate over distance. The perfect example is a volcano eruption, which has a well-defined center, and lava piling up and dissipating over hundreds or thousands of meters to create a peak as in Fig. 8.1.

This dissipation idea is very general. For example, we might model the spread of take-up of a new electric scooter vehicle – such as the DriveDaddy *Rolly* – amongst the population of a city in a similar way. One early adopter will buy the scooter, then will move around the city telling her friends about it. Their friends tell friends-of-friends and so on, and the idea dissipates around the city. If we assume that friends tend to live near each other, that the population and friendship links have homogeneous density, and every friendship has an equal chance of converting into a purchase, then we will see the same volcano-shaped function if we plot take-up over space. Almost any physical or human map we plot will show some level of spatial smoothness like this. In physics, most things tend to dissipate three-dimensionally with spherical symmetry, giving rise to $1/r^2$ flux-conservation laws. In geography, dissipation is usually across two-dimensional domains, where $1/r$ is the equivalent flux conservation law.

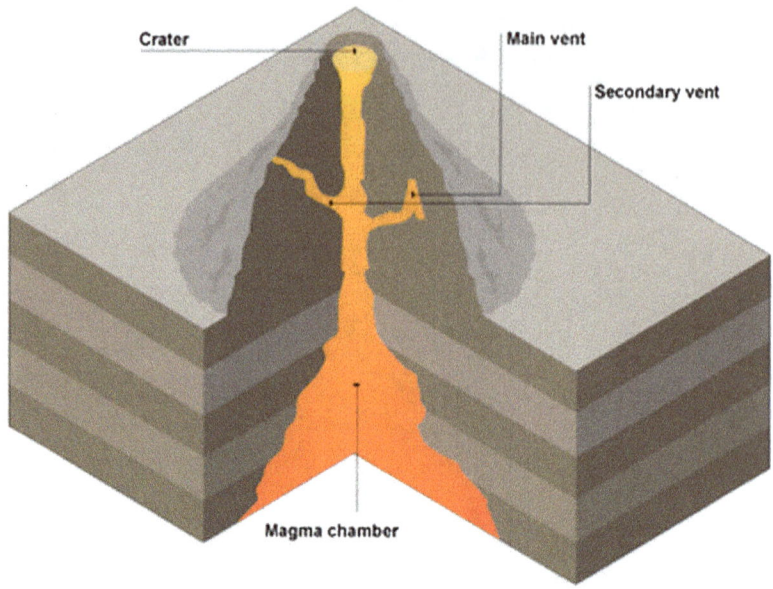

Fig. 8.1 A volcano demonstrates Tobler's first law

8.1 Spatial Statistics

For one-dimensional data such as time-series, we can define an autocorrelation statistic for a signal f_t,

$$R(\tau) = \frac{\langle f_t - \mu \rangle_t \langle f_{t+\tau} - \mu \rangle_t}{\sigma^2}.$$

We can extend this idea to two-dimensions, where we sometimes care about the particular direction of the correlation, which can be described using,

$$R(x', y') = \frac{\langle f_{x,y} - \mu \rangle \langle f_{x+x', y+y'} - \mu \rangle}{\sigma^2}.$$

This might be appropriate, for example, if we are interested in particular directions of traffic flows, or traffic statistics across urban area having specific directional features, such as rivers or hills. In other cases we want to describe spatial correlation independently of direction, which would be more appropriate for transport analysis on a flat, homogeneous terrain. This can be done by averaging over direction-free distances,

$$R(d) = \frac{\langle f_{x,y} - \mu \rangle_{x,y} \langle f_{x+x', y+y'} - \mu \rangle_{x,y}}{\sigma^2}, d = \sqrt{(x'^2 + y'^2)}$$

These will often have the same shape as the volcano function, such as the Radial Basis Function shown in Fig. 8.2. If they are very different from the volcano function then they could suggest that something interesting and unusual is occurring in your data.

By themselves, these are classical (non-Bayesian) statistics, i.e. functions of some data which may or may not be interesting. The spatial R statistics give indications, to a human reader, of how similar the state of f at location x, y is to its state at a near or far away location, $(x + \Delta x, y + \Delta y)$, which is scaled to always take values between $+s1$ (very similar, identical) and -1 (very different, entirely

8.1 Spatial Statistics

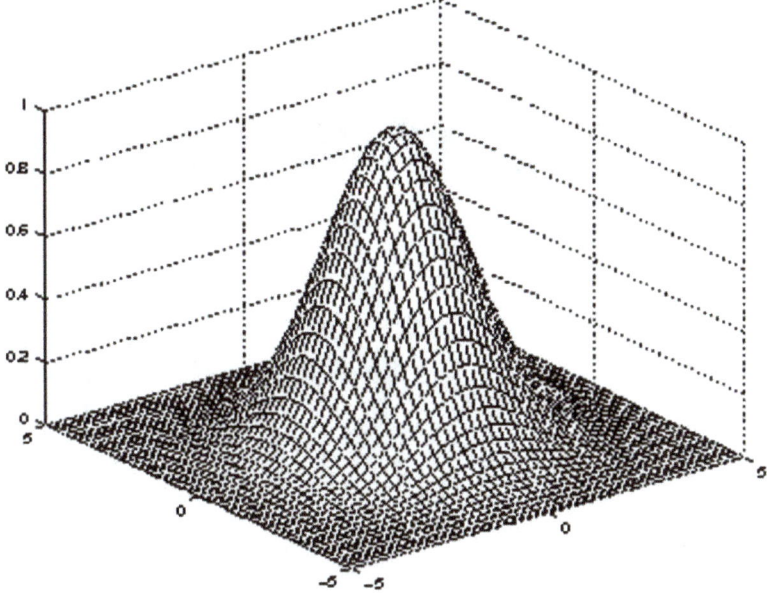

Fig. 8.2 An abstract Radial Basis Function resembling a volcano shape

opposite, but completely predictive) and, with 0 meaning no (linear) relationship. (A non-Bayesian will probably use $\bar{x}$ in place of μ, then run various confidence tests to give more hints about the meaning of R.)

8.2 Bayesian Spatial Models

Bayesians view spatial analysis from a more generative viewpoint, inferring generative parameters instead of inventing and calculating descriptive statistics. There are many possible topologies for spatial Bayesian network models, from which we will sample a few below.

8.2.1 Markov Random Fields (MRF)

The underlying causal processes beneath geographic observations generally have forms like the volcano example. Some unobservable spatial events $f_{x,y}$, give rise to observable data $g_{x,y}$. The underlying causes might in some cases be spatially independent, for example a volcano eruption is (under some assumptions and scales) a single unique, spatially localized event which a priori is uncorrelated with probabilities of similar events occurring nearby. But typically such an event at x, y will give rise to effects $g_{x+x',y+y'}$ at many locations around x, y as well as at $g_{x,y}$ itself. If we discretize space into a regular grid, and assume that these effects are limited to neighboring cells, then we can represent this process as a Bayesian Network as in Fig. 8.3.

To model more distant spatial effects, we could add additional arrows from $f_{x,y}$ to $g_{x+n,y+m}$. In Fig. 8.3, we have $\{n, m\} \in [-1, 0, +1]$ but these could be extended to larger distances. The weights on these causal links would play similar roles to the R autocorrelation statistics of classical analysis, but are explicitly generative and causal in the Bayesian model. We may call this a Spatial Bayesian Network. (Or a Spatial *Causal* Bayesian Network if it also respects causal semantics.)

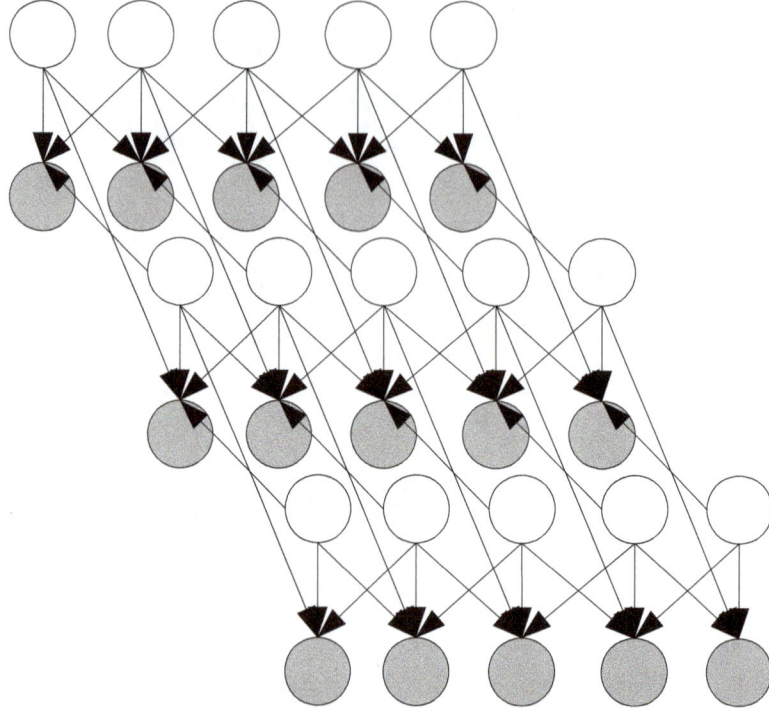

Fig. 8.3 Spatial Bayesian network

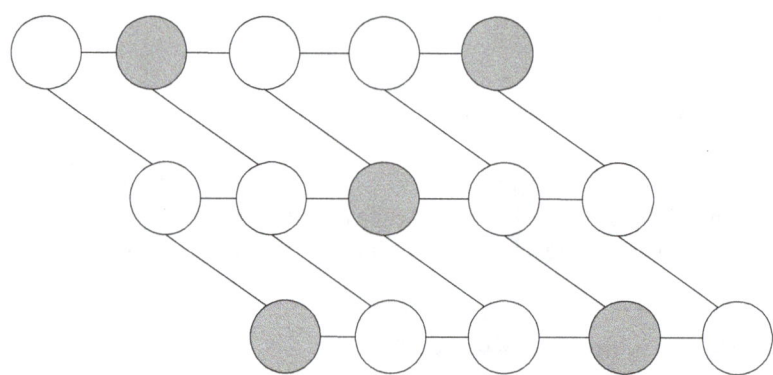

Fig. 8.4 Markov Random Field

In practice, inference in this type of network is computationally complex (as can be guessed from the complexity of even drawing its graph for Fig. 8.3 in a drawing program) and it is common to use various types of simplified model to aid computation. A common reduction is to replace causal networks with correlation-based models. Graphically, this can be done with a single layer of nodes over 2D space, as in Fig. 8.4, where the undirected (not having arrows) links formally represent probability distributions over g_{xy} directly as,

$$P(\{g_{xy}\}_{xy}) = \frac{1}{Z} \prod_{xy} \prod_{\{x',y'\} \in links(x,y)} \phi(g_{xy}, g_{x',y'}),$$

where $links(x, y)$ is the set of neighbors of node x, y defined by the graph, ϕ are "potential factors" (or just "potentials") generating pair-wise correlations, which could include, but are not limited to,

8.2 Bayesian Spatial Models

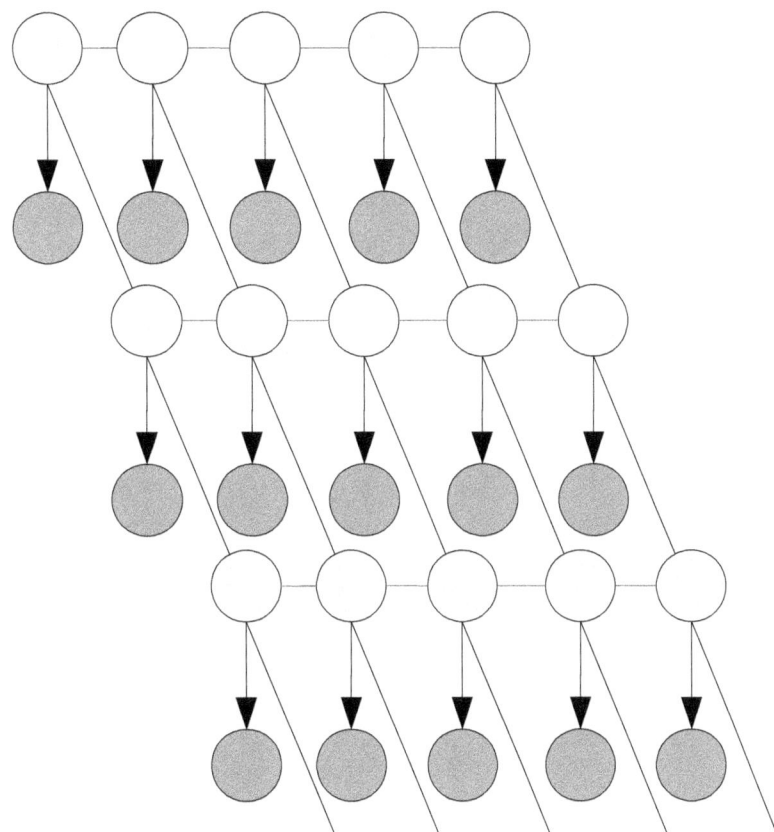

Fig. 8.5 Hidden Markov Random Field

classical R spatial-autocorrelation functions; and Z is a normalizing term such that P sums to 1. This model is known as a Markov Random Field. In the special case where the ϕ factors are homogeneous over space and a regular spatial grid is used, the MRF special case is also known as an "Ising model". These models can be generated in PyMC3 using "potential nodes". As with the Spatial Causal Bayesian Network, we may model longer-range spatial correlations by adding addition links and potential factors between more distant nodes.

For geographic data, we often have observations from some nodes of an MRF and wish to infer values of others, as shown by the gray (observed) and white (unobserved) nodes in the above figure. Inference is performed automatically by software like PyMC3 in this case, and can output probabilities of values of individual or groups of unobserved nodes.[1]

It is possible to combine directed Bayesian Network and undirected factor model elements into models which are given the general name "graphical models" and to define probability distributions of the form,

---

[1] Most spatial models, by virtue of being two-dimensional, are inherently computationally $\mathcal{NP}$-hard to compute. This arises from the presence of many loops in their graphs. Once hardness is present, it is necessary to use approximation algorithms for the Bayesian computations, such as PyMC3's use of MCMC sampling. This is not the case for one-dimensional equivalents, which are known as Hidden Markov Models (HMMs). Much faster, real-time analytic algorithms are available for HMMs and similar models, which are used extensively for time-series data, most famously in real-time speech recognition. As this course emphasises transport data, we generally work with MCMC throughout rather than study these relatively rare, special cases of Bayesian inference.

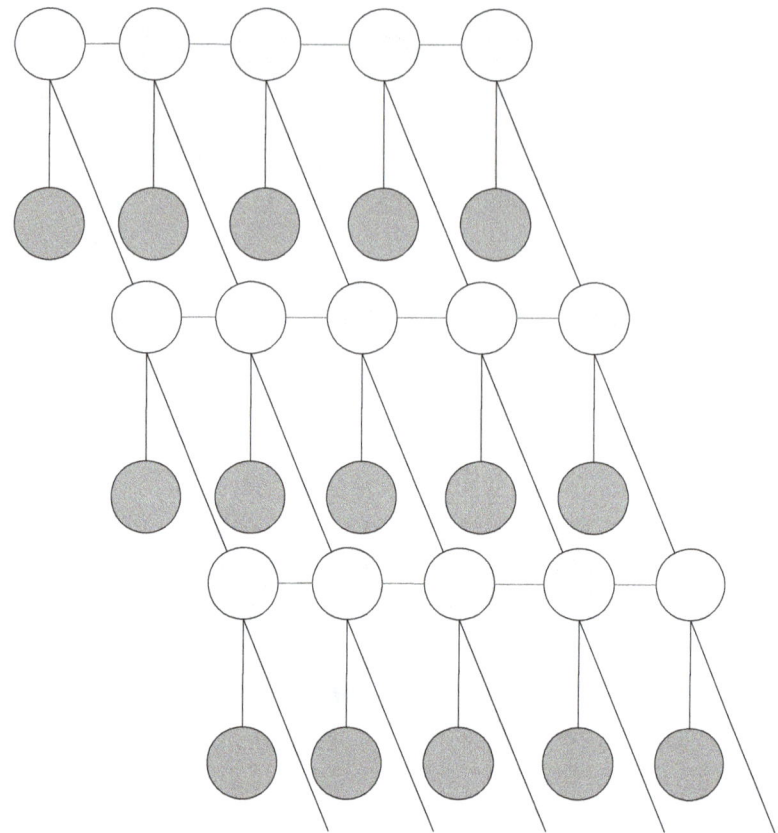

Fig. 8.6 Conditional Markov Random Field

$$P(X_1, X_2...X_M) = \frac{1}{Z}\left(\prod_i P(X_i|\{X_i^{(1)}, ..., X_i^{(N_i)}\})\right)\left(\prod_i \prod_{i' \in links(i)} \phi(X_i, X_i)\right),$$

where *links* again range over undirected links, and the $P(X_i|\{X_i^{(1)}, ..., X_i^{(N_i)}\})$ terms represent the directed links as in standard Bayesian Networks. This equation is for the general case where the index i ranges over all nodes in an arbitrary network. For regular grid-based spatial models we usually use two-dimensional indices, taking $i = (x, y)$.

Other variations of the MRF model include Hidden Markov Random Fields (HMRF), which assume that observations are independent of one another given a state of a spatially correlated unobservable process, as in Fig. 8.5. Also, Conditional Markov Random Fields (CRF), which are like HMRF but replace the causal model of the observation with a simpler correlation, which can lead to faster computation, as in Fig. 8.6.

All of the models using undirected links are approximations to the true causal, generative processes, and are used to make computation easier. As with choosing classifiers, you can often try them all and use the one that happens to work the best rather than worrying about exactly what approximation assumptions are allowable. If it works, use it; if it doesn't work, don't. For various big MRF-type models, specialist software exists to exploit properties of the specific topologies, to go faster than the generic PyMC3.

8.2 Bayesian Spatial Models

Fig. 8.7 MRF for UK counties

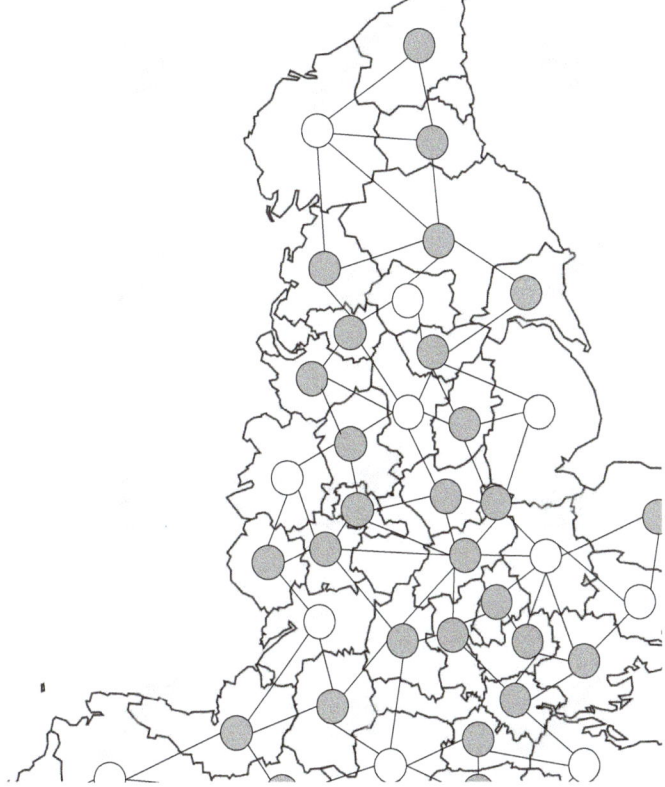

The node topology used in MRF type models does not have to be regular, and for geographic data it often makes sense to treat existing defined regions as nodes in the network. For example, the fragment of a UK map shown in Fig. 8.7 has one node per county, which would be useful if data has been collected at county level. This particular network is an MRF with some observed and some missing data, assuming correlations between neighboring counties only.

Where we have a mixture of observed and missing data like this, we can estimate the correlation potentials from the set of pairs of neighboring counties which both have data, then use them with the network topology to compute posterior beliefs about the missing counties.

8.2.2 Gaussian Processes (Kriging)

Rather than model space as a discrete graph, an alternative is to treat it as continuous. Suppose we have numerical observations from four (gray) points located in a physical 2D space, as shown in Fig. 8.8 left, for example of terrain height or air quality, and are interested in the probable value at an arbitrary unobserved (white) point.

If we know the spatial correlation as a general function of distance $R(d)$, usually having a volcano-like form due to the First Law of Geography, then we may instantiate an MRF on the fly, including the four observations as nodes and the arbitrary point as an unobserved node this graph as a region MRF,

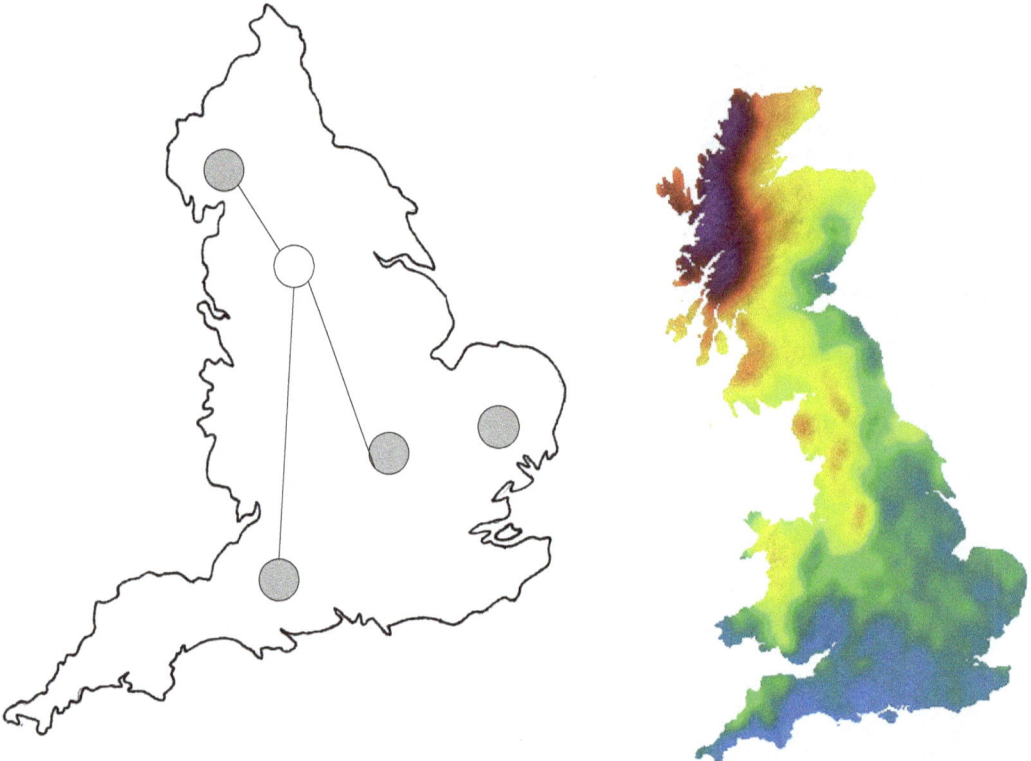

Fig. 8.8 Left: a moving unobserved (white) node can be swept over the continuous map, with factor strengths given by the distance to (four) fixed observed nodes (grey). Right: Example result of inferring pixels across a whole map in this way

with potential factors on the links. We can then move the arbitrary unobserved node around the whole region, for example to each pixel in a map image and compute the posterior belief there. If we plot the mean of these posteriors then we obtain continuous heat maps as shown in Fig. 8.8 right. This map shows only mean values, not the uncertainties. This method is known as Kriging,[2] or more generally (e.g. in different dimensions and as part of larger Bayesian models) as a "Gaussian Process" (GP). Like nearest neighbor classifiers, GPs are "kernel" methods, because they use raw observed data values in their computations of each inference, rather than parameters of any model fitted to them.

8.3 Vehicle Routing

For transport data, we often want to find the shortest or fastest path between points in a road network. This is a computationally "easy" problem, which can be solved quickly using the standard Dijkstra algorithm, and with either road lengths or expected road travel times used as the costs. For example, we might begin with a database of road lengths (as in OpenStreetMap data) and convert these lengths to approximate durations by multiplying by factors for the different highway types such as motorways and country lanes.

---

[2] After D.G. Krige who developed the technique, with G. Matheron, in the 1960s.

8.3 Vehicle Routing

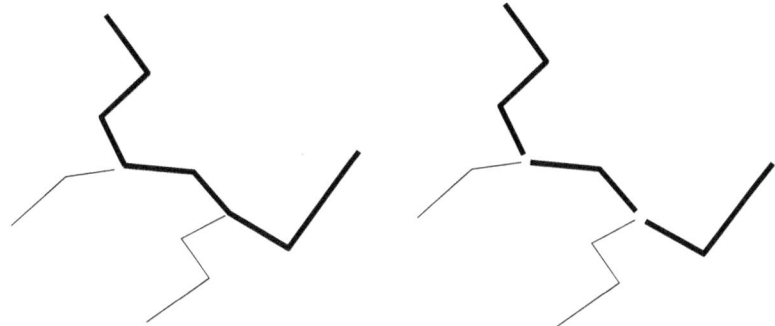

Fig. 8.9 Link-breaking for routing. Left: before link-breaking. Right: after link-breaking

Dijkstra's algorithm is appropriate for sparsely populated transport networks, where the effect of other road users is negligible, or where congestion levels are already known for other vehicles (e.g. when routing a single driver via an in-car navigation system connected to live traffic data). For tasks involving modelling the distribution of traffic as a whole, there is a huge classical modelling literature based around purely mathematical assumptions and models, which makes little use of data. Wardrop's *user equilibrium* names the assumption that each driver knows the position of each other driver on a network, and then choses the fastest route given these congestion levels; while *social equilibrium* is the theoretical optimal set of routes for all drivers, if they could be assigned by a benevolent central authority aiming to minimize the sum of their travel times. These basic models assume each driver has a fixed origin, destination, and departure time, while more complex models can vary preferences of these parameters. Further additions can include probabilistic route choices and models of driver uncertainty about everything above. Yaron Hollander's book *Transport Modelling for a Complete Beginner* provides the definitive introduction to this field. Figuring out how to combine these mathematical based models with Data Science approaches is likely to become a major research field in coming years.

8.3.1 Link-breaking

An important preprocessing step called "link-breaking" is required to prepare most road-map type data for routing computations, including OpenStreetMap data, as illustrated in Fig. 8.9. A major road such as a motorway is typically stored in OGC format as a single long wiggly line. Minor roads connect to it. The minor roads are each stored as single lines. This representation does not tell us directly how the minor road actually connects to the major road. We must break the major road up into smaller links, so that the minor roads connect to the start and end points of the smaller links. This allows the intersections to be represented as nodes in a connectivity network used for routing. Automated tools are available to perform link-breaking.

8.4 Spatial Features

We often want to use spatial information as input to Bayesian models or Discriminative classifiers. Unlike normal data, spatial data does not arrive prepackaged into vectors of features to input to these models. Rather, it consists of entities and shapes at locations. To make it usable for modeling and classification, we need to choose some features to represent it in vector form. There is no rigorous general way to do this. However in practice, common approaches found to be useful include:

- quantize space into regular grid zones, or region zones defined by the data (such as counties),
- classify each zone independently,
- use the presence or absence of entities in a zone as binary features,
- use number of, or area covered by, entities of type X in a zone as real-valued features,
- use "distance to nearest X" as real valued features.

For example, we might have map data containing locations of houses, offices, shops, parks, pedestrian crossings and traffic lights. We could divide our city into 100×100 m square zones (like US city blocks) and count the numbers in each one and/or the distance to the nearest one. We might then use these features to predict output variables such as accident rate or pollution level, or classifications such as residential/industrial, per zone. In some cases, after this step which assumed zones independent of each other, the results could be used as observations in a Hidden MRF model to restore spatial smoothing information.

8.5 Exploratory Analysis

Often we have lots of data and no particular objective other than to "find interesting and useful things". How should we go about doing this?

There is no general algorithm for doing this. In the absence of any theory or model to test against the data, we are first looking for ideas for *what* theories and models might be worth trying out. As such, Data Science is just like real science. Testing Einstein's theory against some data is simple compared with inventing the theory in the first place. Most of what passes for Data Science or "artificial intelligence" only attempts to automate the testing phase and is silent about the process that creates the models.

As in real science, models may be hypothesized by *induction* and *abduction*. In the philosophy of science, these are distinct from each other and from deduction, as:

- *Deduction*: Given A and $A \implies B$, conclude B. Guaranteed to work if the premises are true.
- *Induction*: Given several observations of A followed by B, conclude that $A \implies B$. Not guaranteed to always work, but usually useful in practice. (For example, the world might end after A and before B).
- *Abduction*: Given $A \implies B$ and B, conclude "maybe A".

Abduction is also sometimes known as "priming", and suggests one method for generating hypotheses. It assumes that we know a set of models, from experience in other fields, and some feature in the data "reminds" us one of them, so we get it out and try it. It is a justification for some classical statistics for Bayesians: if they know that some generative model will give rise to certain values of those statistics in data, and they see such statistics, this gives a clue that perhaps that generative model is interesting in the case. Many famous "genius" discoveries come from abduction, where the discoverer has first spent decades internalizing models and theories from many academic and everyday domains, and is reminded of one of them by some superficial feature of data from the problem domain.

Some ways to look for interesting features that might prime models, in order of complexity, include:

- Draw simple plots and histograms of individual variables' distributions. Look for unusual (i.e. non-Gaussian) distributions.
- Plot pairs of variables against each other, look for dependence.
- Compute dimension-reduced projections of many variables via Principal Component Analysis, look for clusters and correlations in them.

8.5 Exploratory Analysis

Fig. 8.10 Snow's cholera map

- Plot variables over geographic space, eyeball them for clusters and correlations.
- Look at statistical correlations and auto-correlations.
- Use discriminative classifiers to see if one variable can be predicted by the other variables (e.g. shopping baskets).
- Use EM algorithms with generative methods to postulate and infer a hidden latent class that generates clusters of data. (This is known as "unsupervised learning", e.g. market segmentation).

If useful predictability appears then it may be worth investigating in more detail, leading to hypothesized generative models. In some cases it is possible to simply search a huge space of possible generative models automatically, without any human theory, just by trying out thousands of Bayesian Network structures randomly or by making small evolutionary changes to them when they predict better.

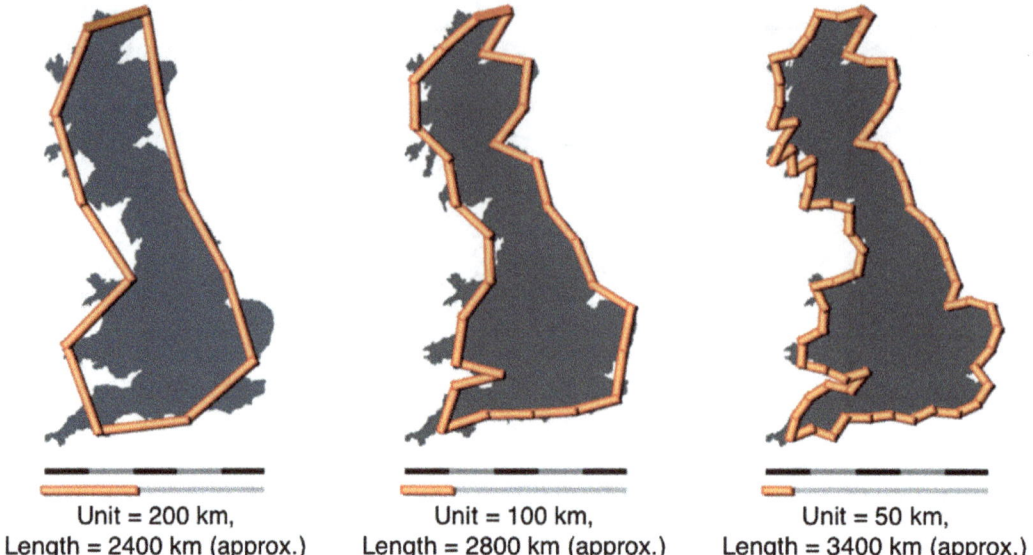

Fig. 8.11 Coastline paradox

A classic example of exploratory data analysis, reproduced in Fig. 8.10, comes from John Snow's 1854 map of Cholera outbreaks in London, which were seen to cluster around wells and pumps, leading to a causal theory that the disease was spread through the water system.

8.6 Scaling Issues

Spatial data is notorious for presenting difficulties regarding issues of scale.

The *"coastline paradox"* problem asks what is the perimeter of Britain. There is no answer to this – any attempt to measure the perimeter is dependent on some choice of smallest straight line segment length, and as there are made smaller, the measured perimeter can increase infinitely, as illustrated in Fig. 8.11. The same problem appears with the measurement of road lengths, for example when using road lengths as costs to compute optimal routes.[3]

Simpson's Paradox. We often want to group data into spatial regions, such as counties or towns. But the choice of how the regions are defined can affect the apparent results. For example, suppose we have surveyed male and female drivers across six UK cities, asking how many have used a mobile phone while driving in the last year. We tabulate the results and get:

	#surveyed	#phone users
Male	8442	**44%**
Female	4321	35%

Colour the whole UK blue for boys, and call the Daily Mail – "UK men are worse phone drivers!". However, if we break the results down by region, to plot on a map, we might see that the same data suggests an entirely different finding:

---

[3] This may have implications for certain travel expense claims.

City	Male		Female	
	Surveyed	Phone users	Surveyed	Phone users
Leeds	825	62%	108	**82%**
London	560	63%	25	**68%**
Manchester	325	**37%**	593	34%
Birmingham	417	33%	375	**35%**
Newcastle	191	**28%**	393	24%
Liverpool	373	6%	341	**7%**

Which would colour most regions in pink for girls. If a newspaper or political party wants to "spin" a story about either gender being the worst offender, they can pick between these two scales of presentation to offer whichever support they need.

Extreme Estimators. In 1999, it was known that the highest cancer rates occurred mostly in rural USA states, and many theories were put forward to explain this. Were these states poorer and lacking medical facilities? Or did they have more outdoor "redneck" workers with skin exposed to sunlight? The mystery deepened when it was also found that the *lowest* cancer rates also occur in rural states! Again, new theories were offered such as rural states having more varied types of workers, or wide mixtures of rich retirees and poorer working people. All of this theorizing was later found to have been a waste of time. The true cause of the apparent result was due to a scaling issue. Assuming each state has some true, unobservable probability of an arbitrary person getting cancer, then the number of cases we observe each year is a frequentist estimator of this generative parameter. Large rural states have lower populations than urban states, so we have fewer observations of cancer and of non-cancer in them. Therefore, there is more uncertainty in the estimates from large rural states than for cities. If we have many of each type of state, then it is more likely that statistical noise will yield extreme values of both types in these states, saying nothing about the underlying generative parameter. (See Gelman, 1999).

8.7 Exercises

8.7.1 Gaussian Processes in GPy

Here is how to work with Gaussian Processes in Python, using the GPy library. It is easier to see what is going on in a one-, rather than two-, dimensional example at first. Here we generate 5 data points from a known sine function, define a correlation function (known as "the kernel") as a Radial Basis Function (i.e. similar to the volcano shape), and plot the mean and standard deviation of posterior beliefs for other nearby values as the center and thickness of a colored band,

```
import GPy, numpy as np
#generate some test data
N = 5
X = np.random.uniform(-3.,3.,(N,1))
Y = np.sin(X) + np.random.randn(N,1)*0.05
#fit and display a Gaussian process
kernel = GPy.kern.RBF(input_dim=1, variance=1., lengthscale=1.)
m= GPy.models.GPRegression(X,Y,kernel)
m.optimize()
m.plot()
```

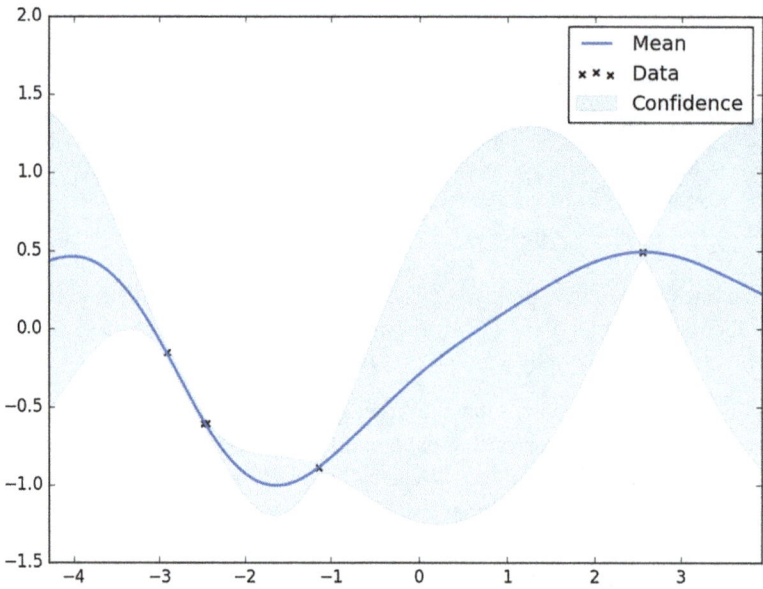

The plot shows that we are more certain in our beliefs when we are near one of the observations than when we are far away. This uncertainty measurement can be useful in optimizing "where to look next" in cases where we can actively collect new data. For example, where to install new traffic or air quality sensors to get the most information. (In the above figure, the best place to look next would be around $x=1$).[4]

We can build a 2D geographical model similarly with GPy as follows,

```
import GPy, numpy as np
#generate some data
X = np.random.uniform(-3.,3.,(50,2))
Y = np.sin(X[:,0:1]) * np.sin(X[:,1:2])+np.random.randn(50,1)*0.05
#define, fit and plot 2D Gaussian Process
ker = GPy.kern.Matern52(2,ARD=True) + GPy.kern.White(2)
m = GPy.models.GPRegression(X,Y,ker)
m.optimize(messages=True,max_f_eval = 1000)
m.plot()
```

---

[4]This can get complicated if you are considering options for locating multiple new sensors at the same time. Usually a reasonably heuristic is the "greedy" method of choosing them in sequence with the best entropy reduction each.

8.7 Exercises

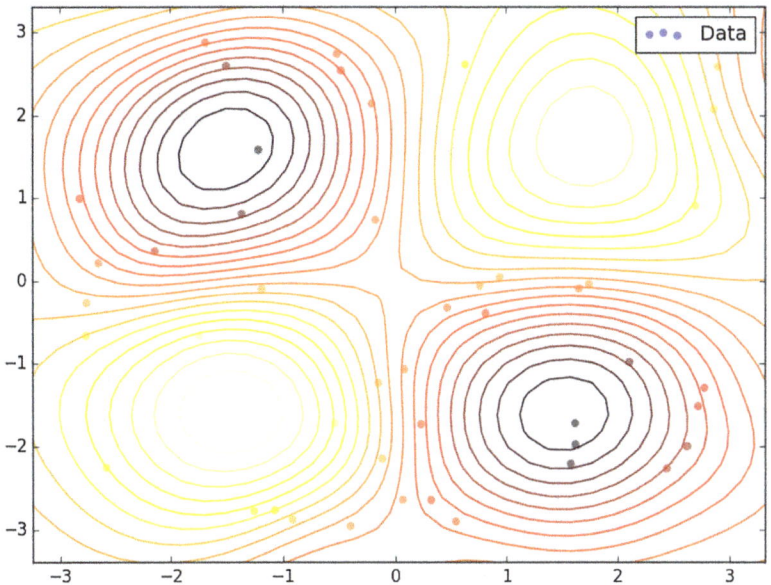

In this case, the plot only shows the mean of the posteriors, as colored contours. If we want to inspect the uncertainties, GPy lets us cut 1D "slices" through this figure, to display their means and standard deviations in full, as follows,

```
slices = [-1, 0, 1.5]
figure = GPy.plotting.plotting_library().figure(3, 1)
for i, y in zip(range(3), slices):
m.plot(figure=figure, fixed_inputs=[(1,y)], row=(i+1), plot_data=False)
```

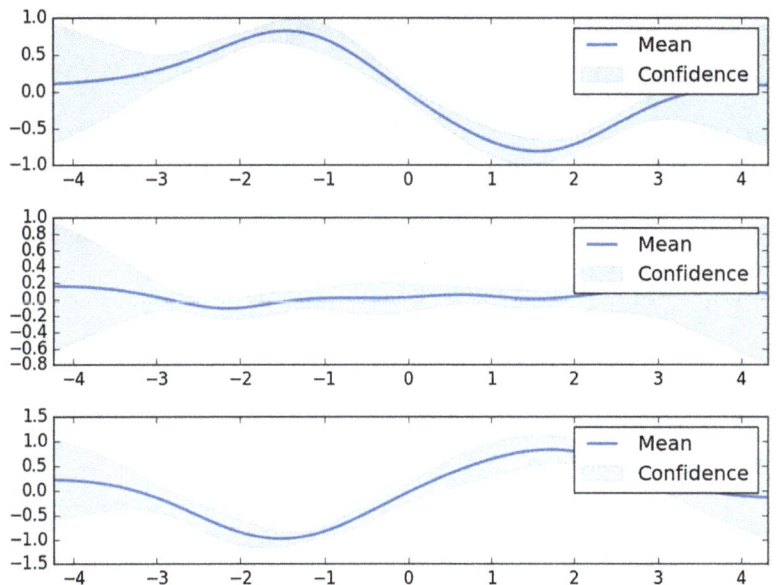

The basic Kriging model assumes that each point can take any real value, but sometimes you may have spatial data which is constrained in someway, such as being all positive, or consisting of probabilities between 0 and 1. In this, and many other, cases, it is possible to use more general Gaussian Process models than Kriging, see the GPy website's tutorials for many examples. (*sheffieldml.github.io/GPy*)

8.7.2 Gaussian Process Traffic Densities

Try to fit and plot a Gaussian Process model to Derbyshire's Bluetooth sensor data, assuming that the county is a homogeneous plane rather than a discrete road network. Under this model, where would be the most useful places to install additional sensors to gather traffic information? In what environments and what conditions might this type of model be useful for transport planning? Is it appropriate for Derbyshire's network?

8.7.3 Vehicle Routing with PostGIS

Here we will make a satnav-like system to route a vehicle through Derbyshire. We must first perform link-breaking, which can be performed for Postgres/PostGIS automatically from OpenStreetMap data with the *osm2pgrouting* tool (*not* from the simple OSM imports used in earlier chapters, as this is a complex process), for example,

```
$ osm2pgrouting -f ~/data/dcc.osm -d mydatabasename -U root -W root
```

(An *.osm* file is also included in the Docker *data* folder.) Use *psql* to examine the many new tables created automatically in the database by this tool.

Dijkstra's algorithm is built into the Postgres extension *pgrouting*, which must be enabled in *psql* with,

```
CREATE EXTENSION pgrouting;
```

This provides extra SQL commands such as *pgr_dijkstra* which can be called as follows to find the shortest route between two nodes in the routing graph,

```
sql = "SELECT * FROM pgr_dijkstra('SELECT gid AS id, source, target, \
length AS cost FROM ways', %d,%d, directed := false), ways \
WHERE ways.gid=pgr_dijkstra.edge;"%(o_link_gid, d_link_gid)
df_route = gpd.GeoDataFrame.from_postgis(sql,con,geom_col='the_geom')
```

See the sample code *routing.py* which includes plotting of the route as in Fig. 8.12, or try to write such a plotting program yourself. Can you improve the routing by using expected durations of road links in place of physical lengths? (*Hint*: OSM data includes highway types such as "motorway" and "trunk" which can be used in these estimates. For a real system, you would usually also make use of live congestion information here too.)

8.7.4 Finding Roadside Sensor Sites

A related problem to routing is computing the closest road link to point, such as a Bluetooth sensor site, which can be done with *ST_Distance* like this,

```
sql = "SELECT gid,name,ST_Distance(ways.the_geom,ST_SetSRID( \
ST_MakePoint(%f, %f),4326)) FROM ways ORDER BY 3 ASC LIMIT 1;"%(o_lon,o_lat)
```

This is especially useful when we want to know what road each Bluetooth sensor is actually monitoring, so that we can link Bluetooth detections to roads in the network.

8.8 Further Reading

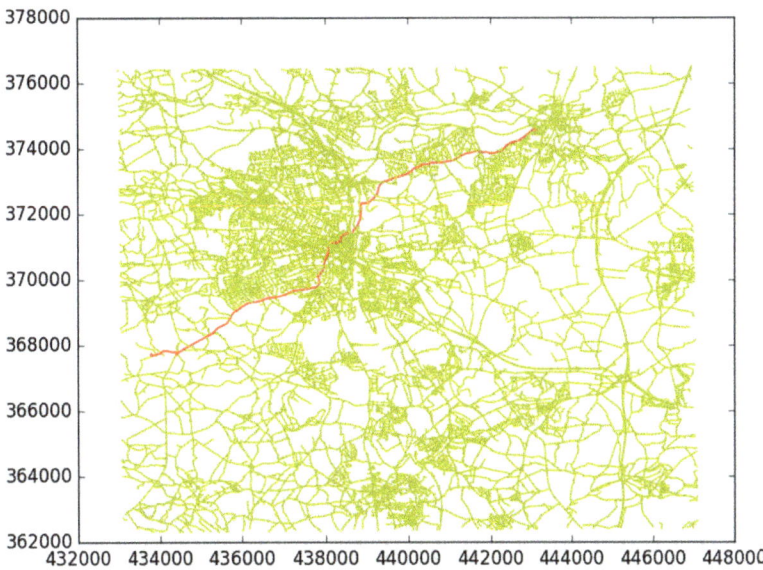

Fig. 8.12 Output of Dijkstra physical distance routing between two points in Derbyshire

8.8 Further Reading

- Koller D, Friedman N (2009) Probabilistic Graphical Models: Principles and Techniques. MIT Press (MRFs and related models.)
- Rasmussen E, Williams C (2006) Gaussian Processes for Machine Learning. MIT Press (Gaussian Processes.)
- http://www.spatial.cs.umn.edu/Book/sdb-chap7.pdf (A draft book chapter on spatial data mining.)
- Hollander Y (2016) Transport Modelling for a Complete Beginner. (Definitive introduction to classical transport modelling.)
- Boden M (2004) The Creative Mind: Myths and Mechanisms. Psychology Press (Discussion of scientific theory creation.)
- Gelman A, Price PN (1999) All maps of parameter estimates are misleading. Stat Med 18(23):1097–2258

Data Visualisation 9

We have spent a long time setting up databases, parsing CSV files, fixing quotation marks and date formats, and learning Bayesian models. Now it's time for the payoff: visualizing the results in full colour! This chapter will give a short overview of relevant human visual perception, present a "gallery" of classic transport-related data visualizations, then show how to produce some of your own.

9.1 Visual Perception

9.1.1 Colours

The human eye does not respond directly to the frequency of light, as the ear does to sound. Rather, it contains four types of receptor, and each type responds strongly to a particular central frequency and less strongly to a wide range of nearby frequencies. Figure 9.1 shows the response curves for the four types.

Red, green and blue cone receptors are concentrated in the center of the retina, while rod receptors cover peripheral vision in monochrome. The rods are the reason that you can sometimes see stars in the "corner of your eye" which disappear when you look directly at them. This creates a complex, compressive relationship between the underlying spectrum and the response. For example, identical perceptual yellow light can be produced either by a lamp emitting a single wavelength of 550 nm, or by two separate sources at 560 nm (red) and 530 nm (green) as in a RGB computer monitor. Physically, these are completely different things.

As we have three colour receptor types, we may consider perceptual colour space as consisting of three dimensions. These dimensions can be used to display information. They are sometimes represented as an RGB (red, green, blue) cube (Fig. 9.2 left), but also in other ways such as Hue, Saturation, Value (HSV) cones (Fig. 9.2 center). RGB can be useful when we wish to visualize three parameters simultaneously using colour. HSV is useful for visualizing in cases where one parameter wraps-around, as it is described using cylindrical (θ, r, h) coordinates. For example, hue can be mapped onto times-of-day as shown in Fig. 9.2 right.

From an artistic perspective, there are theories about which palettes of colours look nice together, which can be found in painting and fashion textbooks, such as Fig. 9.3's season-based model by a fashion consultancy. Companies like Apple take palette design very seriously, enforcing certain systems on app designers to create a consistent brand. We won't go into these theories here other than to note a

© Springer International Publishing AG 2018
C. Fox, *Data Science for Transport*, Springer Textbooks in Earth Sciences,
Geography and Environment, https://doi.org/10.1007/978-3-319-72953-4_9

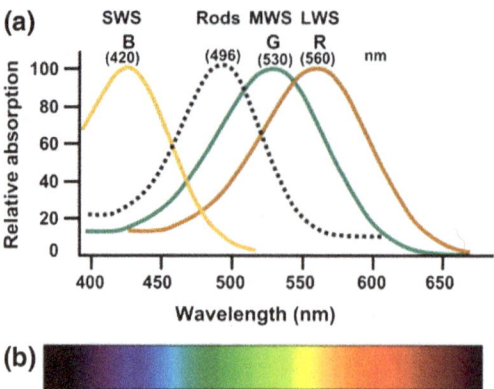

Fig. 9.1 Human eye response to colours. (From *www.sciencedirect.com/topics/page/Cone_cell*)

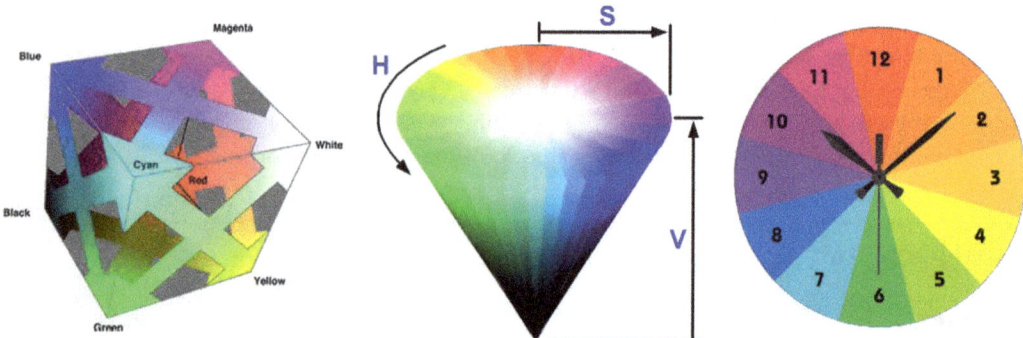

Fig. 9.2 Colour space models

basic principle: that since colour space is three dimensional, it is possible to cut many open or closed, straight or wiggly, 1D lines and 2D surfaces, from the 3D colour spaces, which will all have colour continuity across the parameter(s) but can be chosen to include aesthetically pleasing regions of the spaces. Such lines and surfaces then define coordinate systems which can be used to display 1D or 2D information by their colours.

Be warned that different countries' traditions can assign completely different colour symbolism to the same colour, or sometimes multiple conflicting meanings in the same country. For example red in the UK means danger or love; in Japan it is associated with courage; in some parts of Africa it means life but in others it means death. Don't present your new road safety improvement results in a death colour!

9.1.2 Visual Attention

The eye does not see whole scenes in detail at once, rather the retina only sees the details of a small "fovea" region at a time, as in Fig. 9.4. Humans usually move this fovea to fixate briefly on a different point, in motions called "saccades". Perception of a scene is not usually perception of the retina's foveal "pixels" but rather of a multi-level, model-based interpretation of objects in the whole scene. This is performed by a hierarchy of visual regions in the brain's cortex, shown in Fig. 9.5, and Fig. 9.6, beginning in the LGN nucleus and going through visual areas V1 to V4, which extract

9.1 Visual Perception

Fig. 9.3 Artistic colour palette. (Source: *House of Colour*.)

Fig. 9.4 Foveal attention

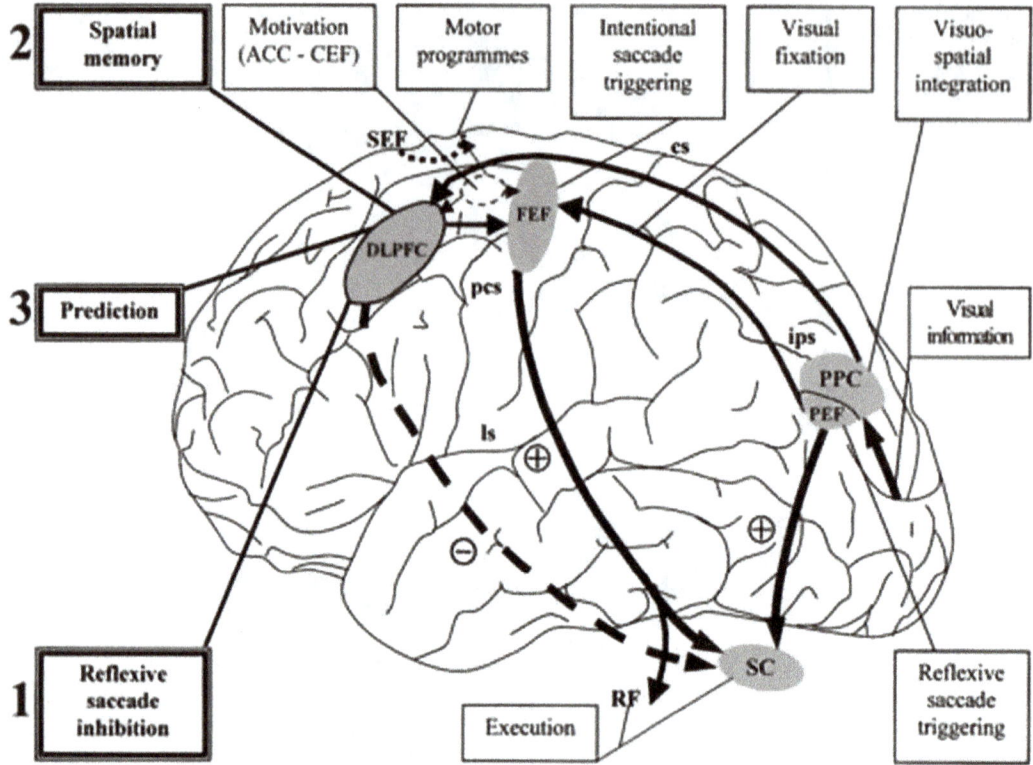

Fig. 9.5 Human visual system anatomy

progressively more complex visual features such as edges, corners, objects, faces, cars, and even (apparently) the actress Halle Berry.[1] Your conscious perception is usually of the highest level of these entities, represented in the highest of the neural areas – such as perceiving Halle Berry as one "thing" rather as a complex collection of pixels, edges or eyes and mouth. But if you concentrate your attention on these details, represented in the lower brain areas, then you can bring them into your conscious perception. You can move this mental "focus of attention" around an image similarly to moving your physical fovea around it. Perception of images including data visualizations is thus an active rather than passive process. Viewers will not usually comprehend your image in a single moment of perception, but will instead explore it over time by moving both their physical fovea and their mental attention around. Like designing a city or a website, you may need to plan for this and consider how to create paths for these explorations over time. Don't throw all your data at the viewer at once – even in a static image you can create signposts and cues for how to navigate through it.

As well as this cortical processing system, there is an independent second visual pathway shown in Figs. 9.5 and 9.6, which connects low-level vision though the Superior Colliculus (SC) directly to the emotional amygdala and to the attention system. This evolutionarily older and computationally faster pathway does not perceive high-level features at all, but simply responds to simple "emergency"-type features of the pixels, such as change of colour areas, and flashes of light. This system responds almost instantly to features such as car brake lights coming on, without requiring the slow computational

---

[1] Quiroga, R.; et al. (2005). *Invariant visual representation by single neurons in the human brain*. Nature. 435 (7045): 1102–1107.

9.1 Visual Perception

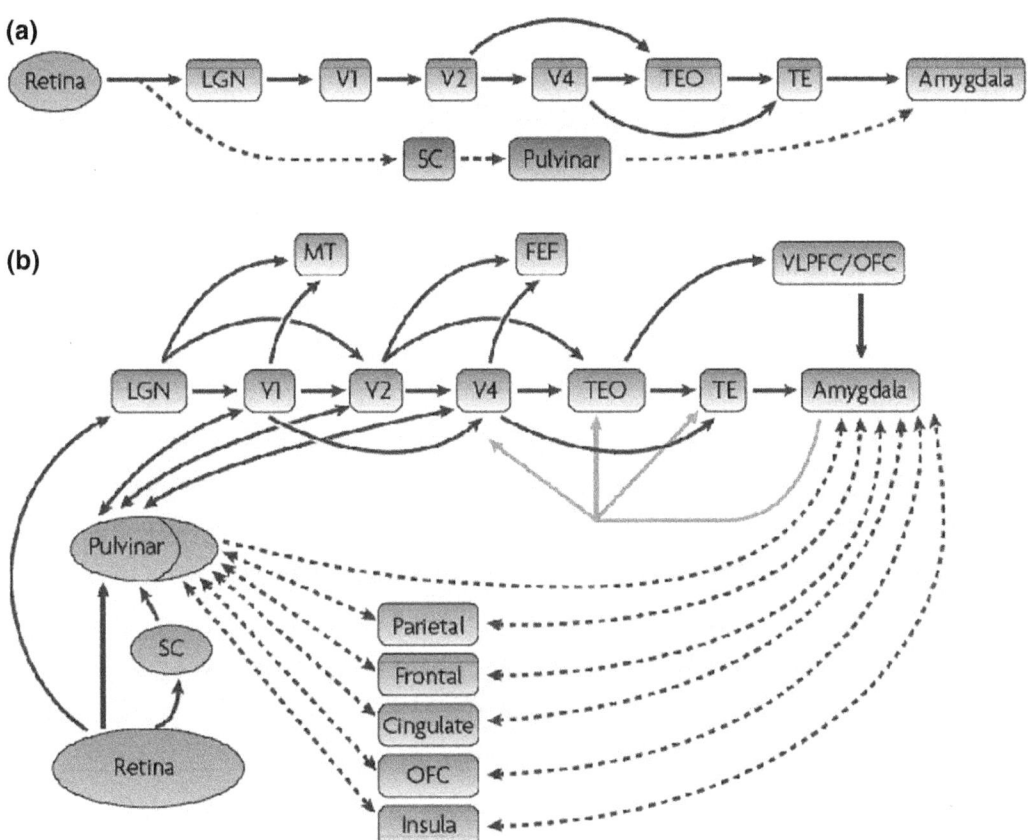

Fig. 9.6 Human visual system computational architecture

process of perceiving the whole car as a "thing" first. Visual attention, both mental and foveal, can be controlled top-down though conscious decisions to explore objects, but can also be overridden by the subconscious SC circuit in such emergencies. You can find the red letter in Fig. 9.7 in $O(1)$ time using your fast, parallel SC system's reaction to the change in colour, but you need to take $O(n)$ time to search for the high-level letter "b" using your slower cortical system. This distinction can be important in visualization because it gives ways to control a viewer's attention rather than have them explore an image in their own way. Knowledge of the SC circuit is also useful in the design of real-time information presentations such as road signs and signal lights, which must quickly bring attention to themselves.[2]

It is widely believed[3] that our perception may contain up to "the magical number seven plus or minus two" distinct objects at any point in time. When more objects are present, we must perceive different subgroups of them in a sequence by moving our attention around them, or use memory techniques such

---

[2]The two systems give rise to hypothesis about a dangerous object at a location in a similar way to how hypotheses are created in Science from the previous chapter. The top-down system requires you to have some prior idea of what is there which directs you to direct attention at a point to investigate it. The SC system uses superficial, bottom-up features of the raw data to make a hypothesis there directly.

[3]Though not yet understood by neuroscience. One current theory is that the seven entities are represented in working memory by discrete phases of an oscillation in the Hippocampus. See Miller et al. 2008, *Single neuron binding properties and the magical number 7*, Hippocampus. 18 (11): 1122–30.

sdfksdfng
zvpdsdvs
sddferdsdi
vudpivdbs
ssdfidysw

Fig. 9.7 Linear and parallel visual search

Fig. 9.8 A "mind map"

as clustering multiple objects together into a single new object. People can get confused and unhappy if you make them do this mental work themselves, but can store very large hierarchies of memories when clustered into entities of this size. The "mind map" concept as illustrated in Fig. 9.8 was designed specifically to exploit this memory effect in visualization.

Humans are especially good at perceiving flat, continuous, two dimensional spaces. So like most visualizations, the mind map is extended over such a space, and makes use of spatial as well as hierarchical relationships for example in the figure, "strategy" and "picture" are drawn close together and linked despite growing from different parts of the hierarchy.

9.2 Geographic Visualization (Maps)

Transport data scientists are fortunate that human perception and visualisation works so well for continuous, flat two dimensional physical space – as used in the mind map – because the data they wish to visualize is often extended over such spaces. A visualization of data in such a space is called a "map"!

As discussed in the previous chapter, estimation of continuous values over space is not trivial, and prone to scaling effects such as choice of region size. Such effects can be exploited to bias the viewer's perception of the data for political purposes. For example some authors favoring Scottish independence from the UK presented the UK as a "divided nation" following its 2016 vote for independence from the EU, using hard classification maps such as Fig. 9.9 (left); while others favoring unity within the UK showed more gradual presentations such as Fig. 9.9 (right).

Another regioning artifact is seen in the travel times map of Fig. 9.10 which appears to take a "nearest neighbor" approach to iso-lines, seen from the blob shapes around individual data points in sparser areas. If all London commuters used this particular visualisation to choose where to live, we would see a massive fall in property values between Isle of Dogs and Greenwich, where the data suddenly "runs out"! We would also expect to see smaller discrete falls in prices between the discretized colored zones imposed by the map's design.

Sometimes we want visualized regions to indicate the size of the population or some other value rather than physical geographical area, but preserving topology as in Fig. 9.11. Again such choices may have a political element: the propaganda value for Scottish independence from the UK provided by these election results may seem less when regions are presented by population rather than geography; but the case for London independence from England may seem greater.

The London Underground map (Figs. 9.12 and 9.13) is a famous example of distorting space to simply the presentation of transport topology. It is loved by millions of people but does sometimes result in pedestrians feeling very disoriented in areas like Westminster (where Whitehall gets rotated a full 90 degrees from reality), or taking unnecessary journeys from Euston Square to Warren St (which are in reality only 180 meters apart).

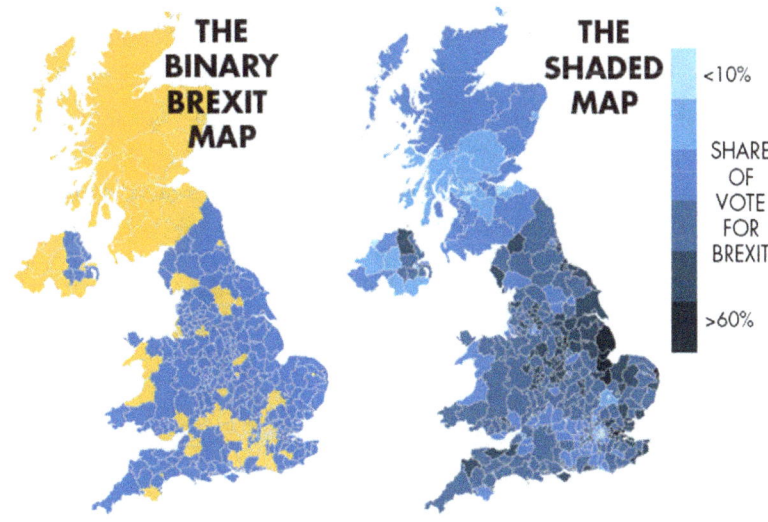

Fig. 9.9 Brexit visualizations. (Source: Fraser Nelson, *blogs.spectator.co.uk/2016/06/sturgeons-opportunity-isnt-brexit-meltdown-pro-remain-unionists/*)

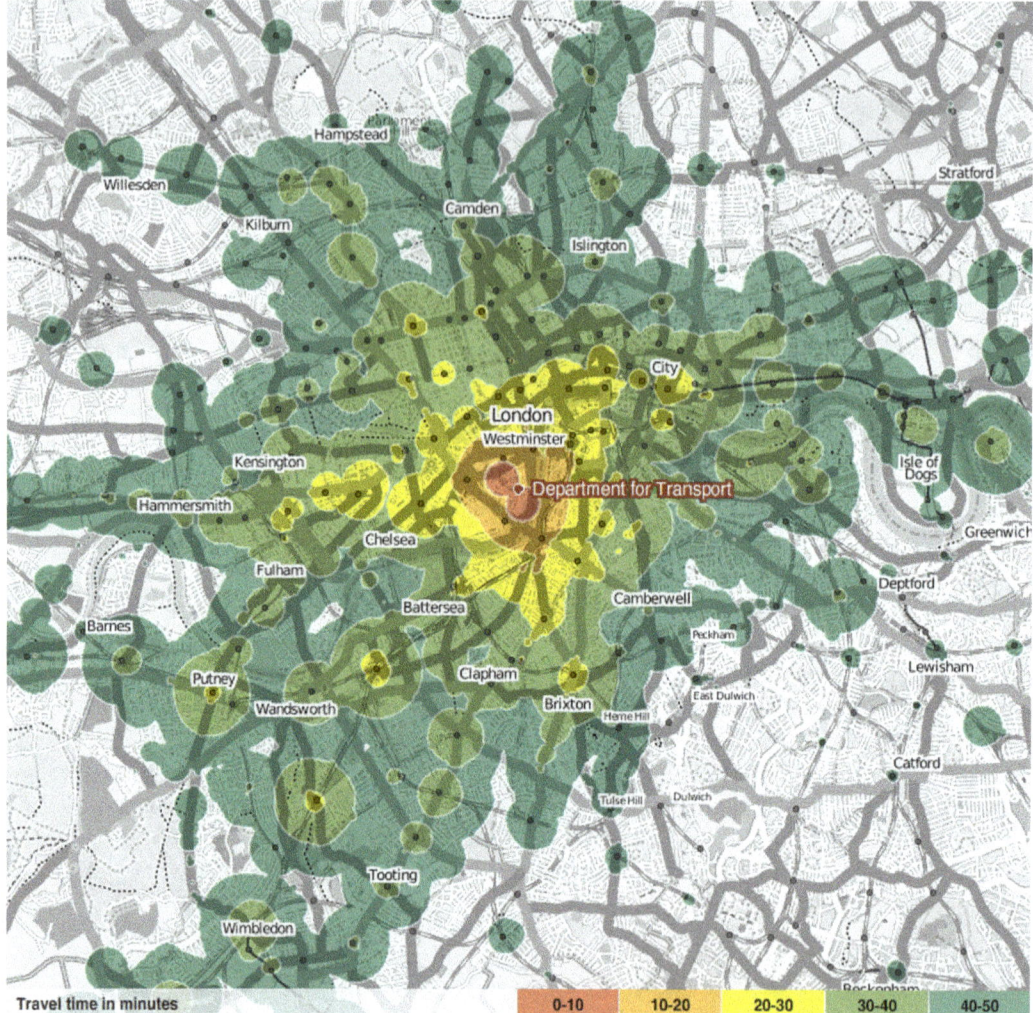

Fig. 9.10 Public transport travel times to reach the Department for Transport in London by 9 a.m. (Source: *Department for Transport*.)

- Are these problems serious enough to be causing loss of income and extra carbon emissions from suboptimal journeys? How could we find out?[4]

In extreme cases maps are designed to deliberately mislead the viewer, such as Fig. 9.14 which uses 3D perspective to apparently relocate a hotel in the "center". More great examples of crimes against cartography may be found in the book "How to Lie With Maps". Don't do this!

---

[4]You are probably already familiar with this map, however every Transport Data Scientist should spend a few minutes of their career admiring it critically!

9.2 Geographic Visualization (Maps)

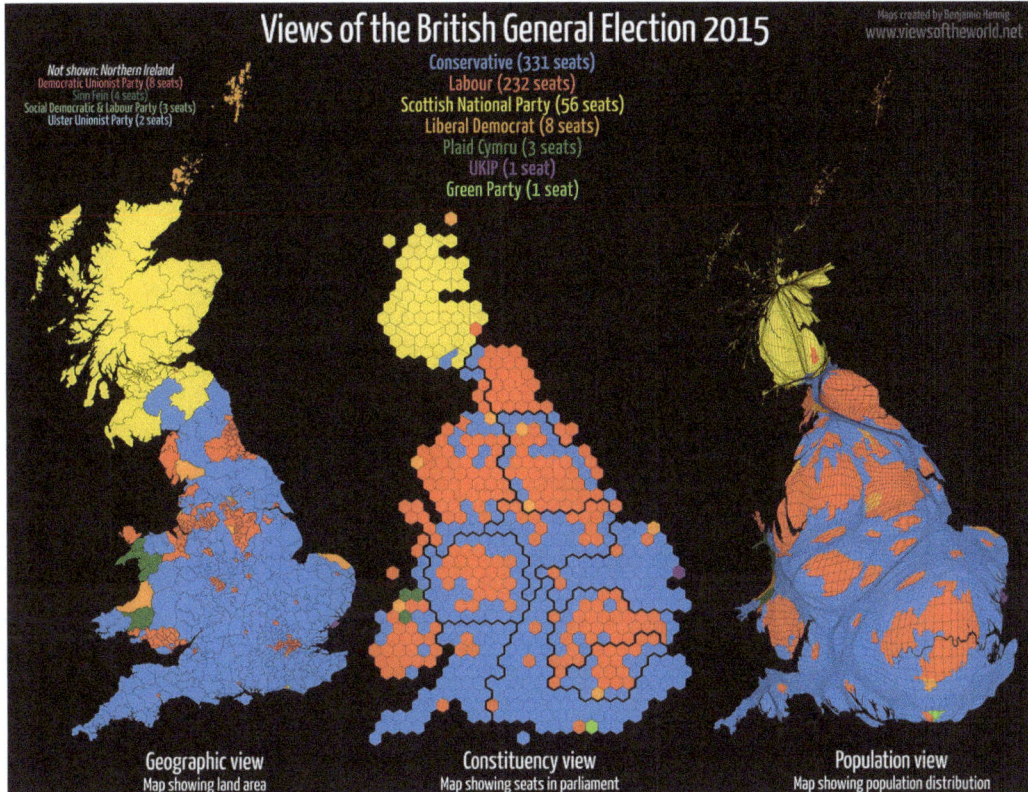

Fig. 9.11 Election maps

9.2.1 Traffic Flow Maps

Minard's 1869 map of Napoleon's campaign of 1812 is a classic infographic, reproduced in Fig. 9.15. It is often considered the best ever, and shows:

- geographical route of the march (x and y axes = longitude and latitude),
- direction of marching (advance = thick light band; and retreat = narrow black band),
- the number of troops (thickness of bands),
- distance traveled (numbers next to bands),
- temperature (lower sub-figure).

Milard's flow thickness concept is generally useful for transportation visualization. For example, Fig. 9.16 is generated from the Derbyshire data. Given a set of origin-destination routes, and counts of those routes from Bluetooth sensors, we approximate the flows through the network by assigning each journey to its shortest path. We then count the total estimated flow on each segment, and display it as a thickness.

134   9 Data Visualisation

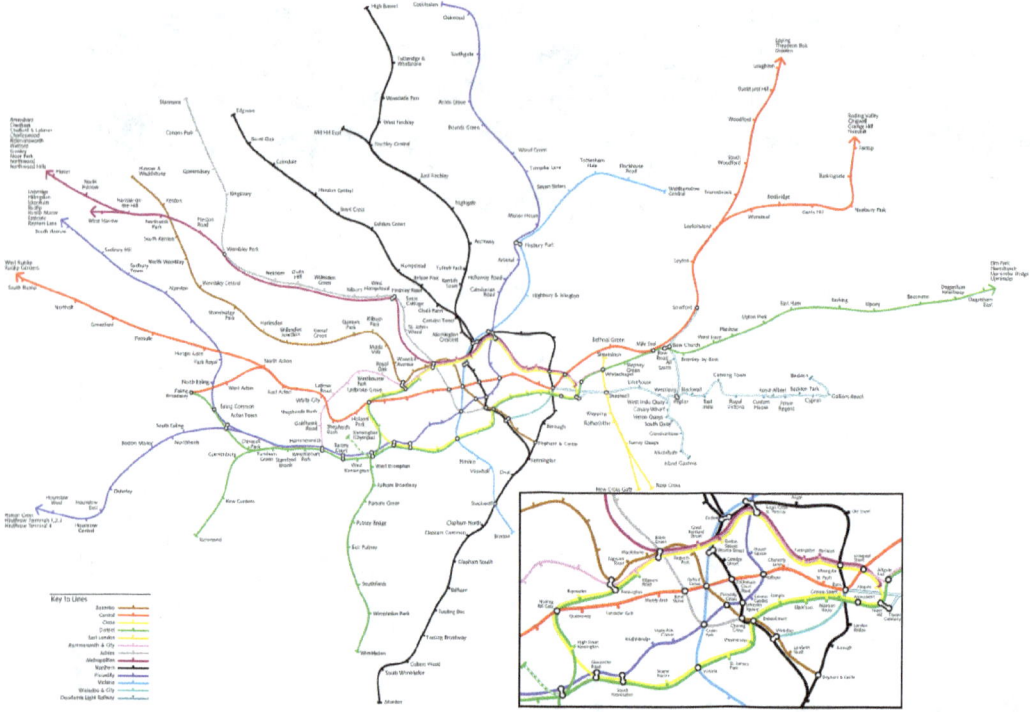

Fig. 9.12 Physical London Underground map

Fig. 9.13 Stylized London Underground map

9.2 Geographic Visualization (Maps)

Fig. 9.14 A misleading map

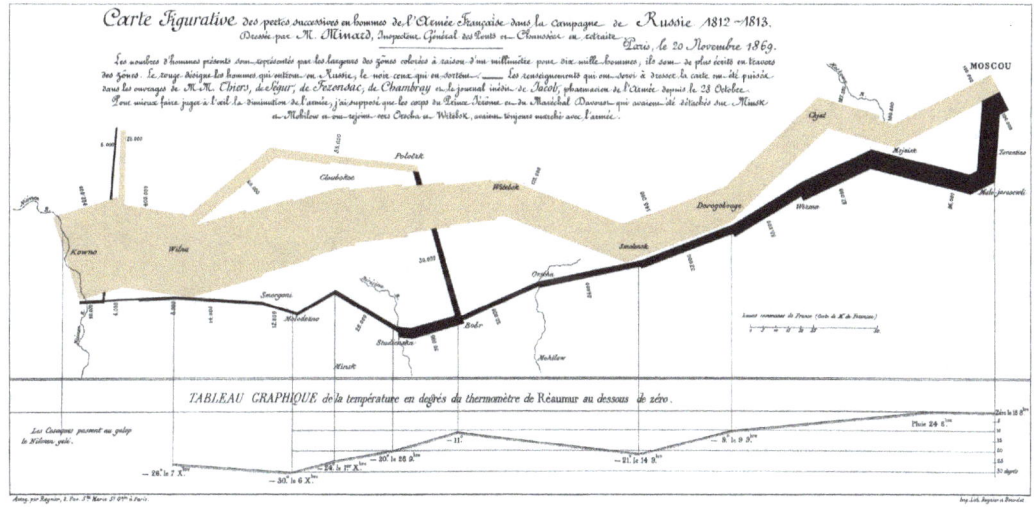

Fig. 9.15 Millard's flow map

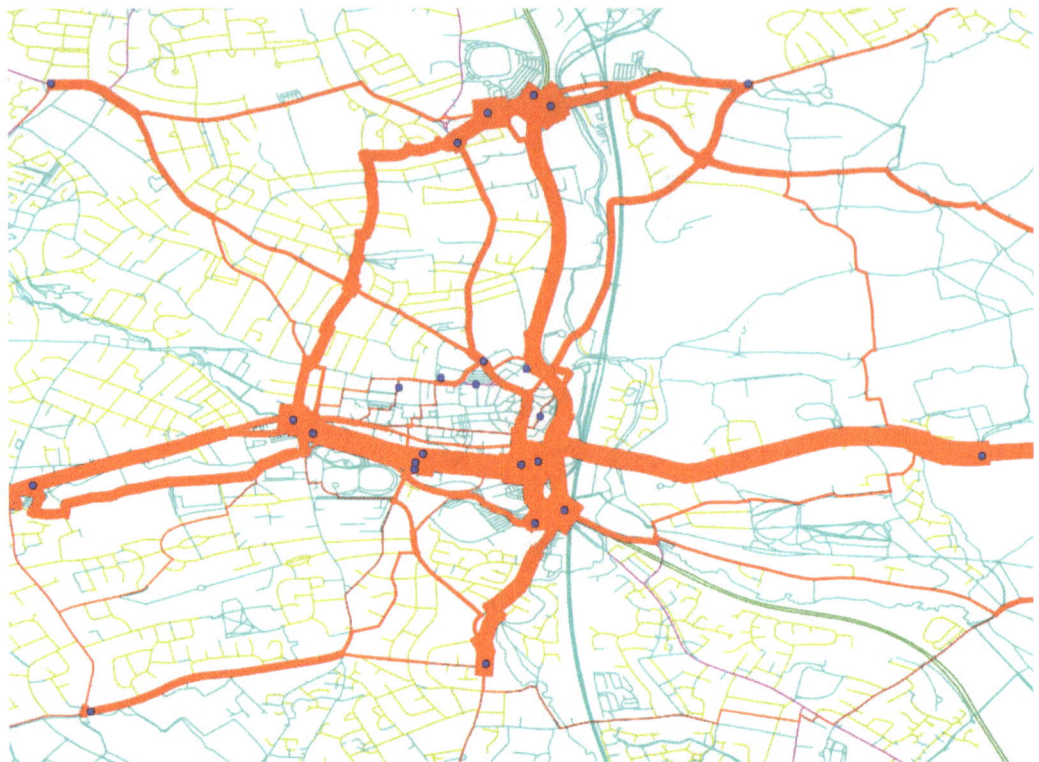

Fig. 9.16 Traffic flows inferred around Derbyshire network from Bluetooth sensors and Dijkstra routing, based on code by ITS Leeds student Lawrence Duncan. Blue circles show sensor locations

9.2.2 Slippy Maps

"Slippy maps" is the generic name given to interactive scrolling and zooming maps that are now common on internet map sites such as *openstreetmap.org*. They are made with various similar programs in a standard way:

- Standard sized images of square "tiles" of maps at multiple scales (e.g. 100 pixels = 100 m, 1 km, 1 km, 10 km, 100 km) are created in advance for every scale and every region of the map, and stored on a web server (e.g. Apache) using standard filenames such as *mymap_scale3_north004667 _east008424.png*, where the filenames include the scale and integer coordinate in the map. The tiles could be produced by basic Python drawing commands as in the flow map above, or a rendering program such as Mapnik could be used to give cartographic quality, as used by OpenStreetMap and shown in Fig. 9.17.
- The user's web browser displays the map web page, which includes a standard embedded JavaScript slippy-map program such as Leaflet (used in OpenStreetMap) or OpenLayers. The slippy-map program is provided with the address of the tile server machine, which is often a different machine from the web server of the map page itself. Slippy-map libraries can also display additional optional overlay layers from Shapefiles or PostGIS, via interfaces such as GeoServer and GeoJSON files. They work by downloading and preparing tiles from nearby areas, ideally trying to predict where the user will want to look next so that the data is already available on the local computer by the time they move.

9.2 Geographic Visualization (Maps)

Fig. 9.17 A web-based Leaflet slippy map rendering of Mapnik tiles, from *openstreetmap.org*

9.2.3 Info-Graphics

Infographic panels like Fig. 9.18 have become common in newspapers and magazines as part of the "data journalism" movement. "Dashboards" (Fig. 9.19) are closely related to infographic panels, but are live systems usually displayed on a web page (or sometimes on printed daily reports) to management Data-focused managers want to see everything important about their complex organizations in one, or a few, images, to get an overview of what is going on. Often dashboards include the ability to "drill down" to explore data in more detail. Some transport authorities are now building whole control center rooms of dashboards to provide "total data immersion" for controllers. Some dashboards also assign numerical scores to management objectives (e.g. travel delays, throughput, pollution) and a single numerical sum score, called "balanced scorecards". These sometimes cause controversy when they lead to their staff "gaming" the metrics to achieve high scores at the expense of delivering real value. Like designing examination mark schemes, the skill here is to choose metrics which more accurately represent what is actually wanted.

Fig. 9.18 An infographic. (Source: Transport Scotland, www.transport.gov.scot/statistics/scottish-transport-statistics-2014-infographics-6498)

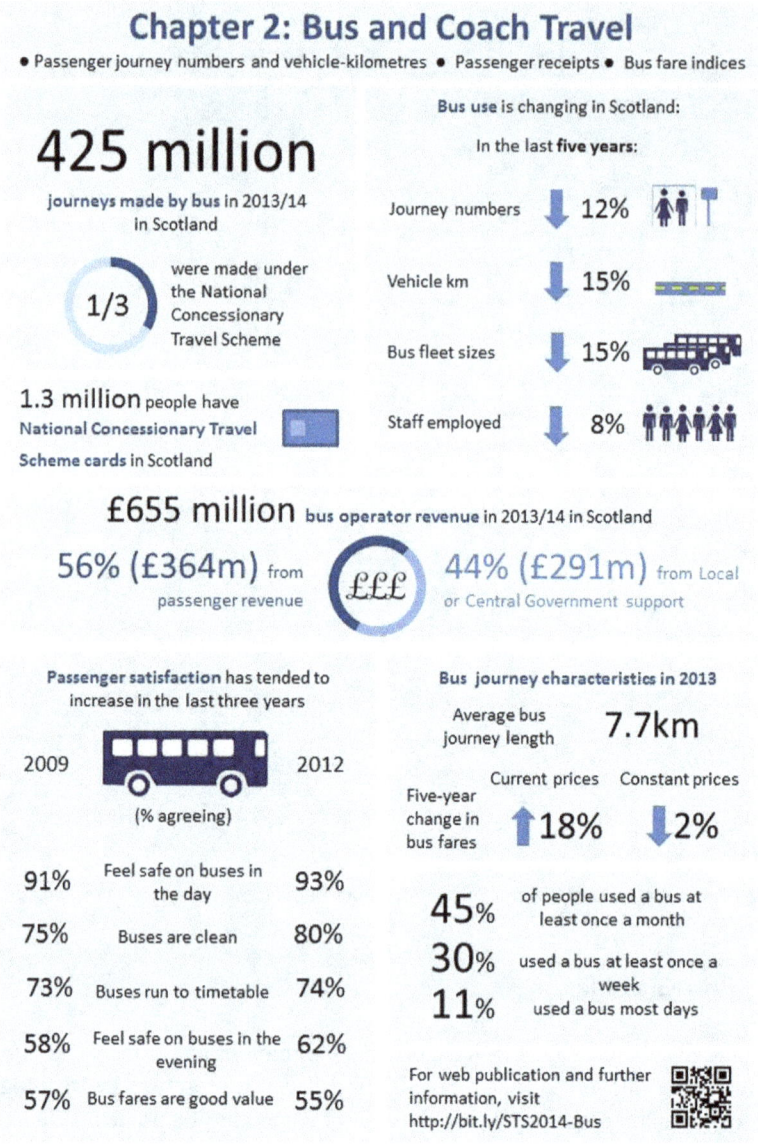

Chernoff faces are an influential type of infographic using simple face shapes with many distinguishable parameters which enable multi-dimensional data to be displayed across space, shown in Fig. 9.20. They have inspired many similar visualizations, for example you might use images of vehicles with different sizes front and back wheels, numbers of passengers, and colours, to visualize many dimensions of transport data. As always, be careful not to subconsciously bias your presentation, for example using bigger smiles to show number of voters for one political party is unlikely to be received well by its opponents' supporters.

9.2 Geographic Visualization (Maps)

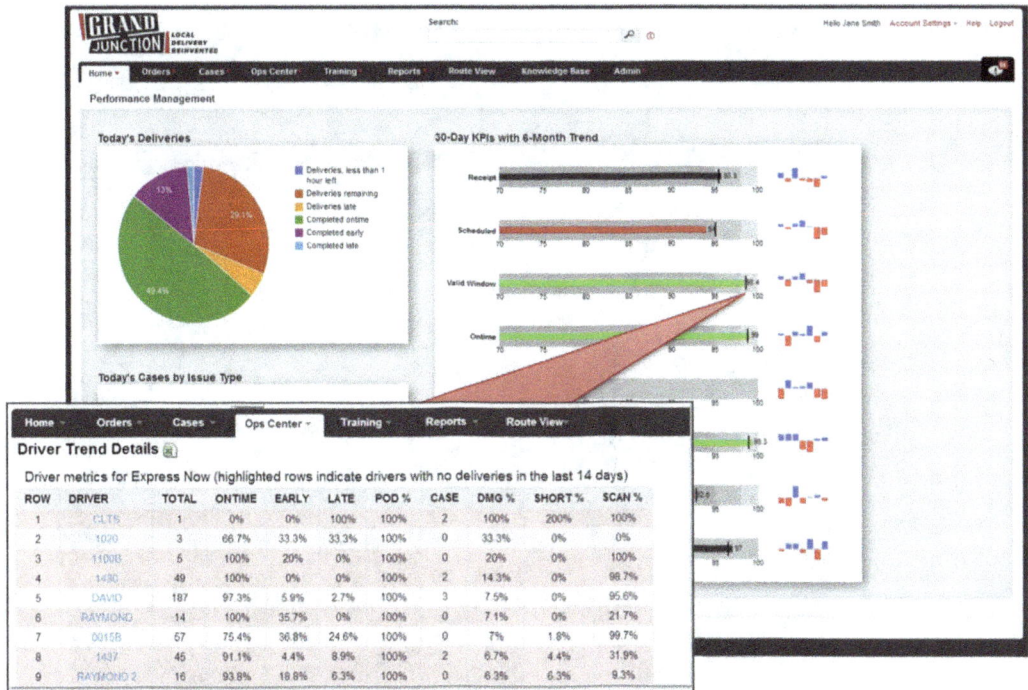

Fig. 9.19 A freight delivery dashboard, with drill-down. (Source: *grandjunctioninc.com*)

A current research area looks at how to display data to drivers in real time through Head Up Displays (HUDs) as in Fig. 9.21. ITS Leeds uses driver simulation and psychology lab studies to evaluate new interfaces. Data can be useful to improve safety and efficiency. For example, a car connected to the road network's ANPR system could look up the licence plates of the vehicles around it, assign them Antisocial Unpleasant Driver Indicator scores according to their driving histories as logged by the rest of the network, and display advance warnings of problem vehicles approaching even before visual contact is made. However, HUDs present additional visual work to the driver, and if not designed carefully can lead to visual distraction from or obstruction of other dangers that are visually present. Pop-out effects are especially important here in grabbing attention; also the human eye needs time to refocus on objects at different depths, unless the HUD can be presented in virtual 3D in the scene. Research might find that HUD distraction effects require real-time driving data to be presented in ways other than visualization – for example, using speech synthesis or other sounds, or tactile feedback such as seat or steering wheel vibration warnings.

Current research looks at how autonomous vehicles should communicate information from data to road users. How can current displays such as indicator and brake lights, and headlight-flashing, and eye contact, be made appropriate for self-driving cars? This is especially important in conflict situations such as negotiating roundabout and slip-road entry, where human drivers would communicate through complex eye-contacts and gestures. Current ideas include the use of emoticon products as in Fig. 9.22 to flash to other road users, and predicting their reactions to them from historical logs of their road behavior.

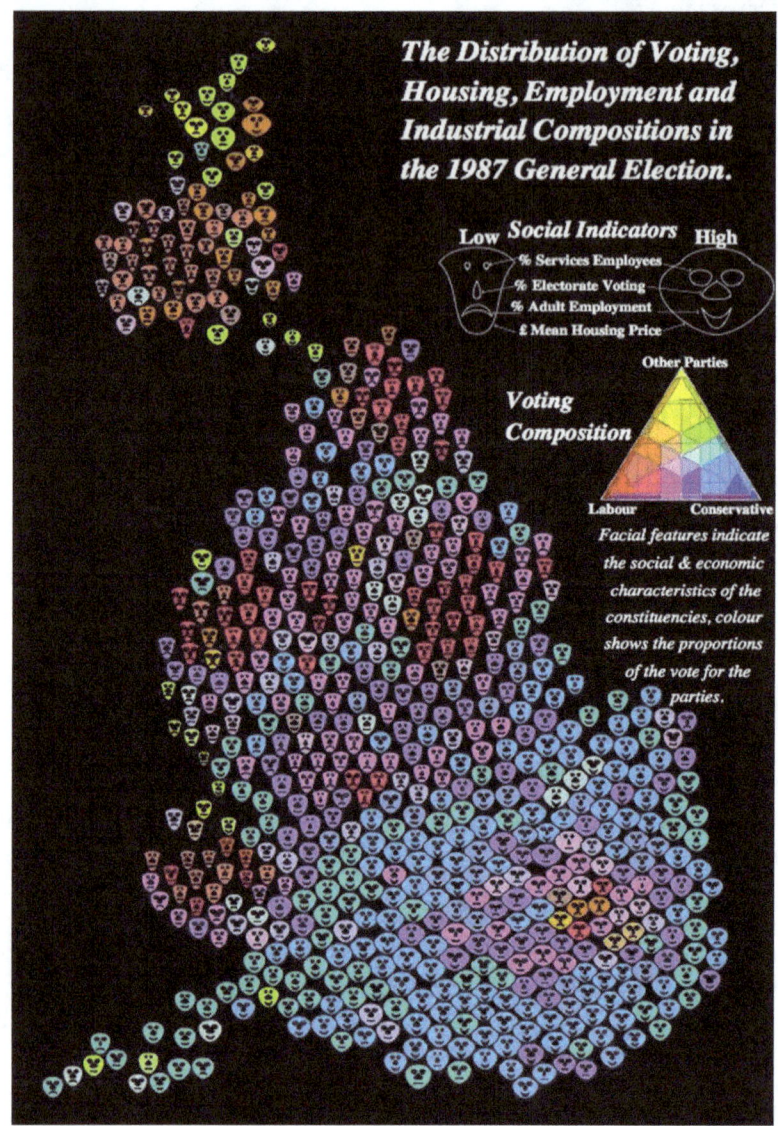

Fig. 9.20 Chernoff faces. Election map by Danny Dorling, 1991

Fig. 9.21 In-car Head-Up Display. (Source: *www.bmwblog.com*)

Fig. 9.22 Current research considers the use of novel signals for manual and autonomous vehicles such as this *drivemocion.com* product

9.3 Exercises

9.3.1 Web Page Maps with Leaflet

Leaflet is a slippy-map tool which allows you to easily draw layers over existing maps, such as OpenStreetMap. It is especially useful for quickly publishing transport data as web pages as it allows you to map use of OpenSteetMap's existing professional graphics for the basic map, and add objects such as markers, arrows and text, and runs inside a web browser. For example, local authorities can use it to quickly provide public transport or congestion maps to their citizens with just a few minutes of coding. In this example we will add a single rectangle annotation showing the location of Chesterfield city center to a map of Derbyshire, and display it in a web page.

Here is some code creating a shapefile containing a polygon defining Chesterfield town center, using the raw OGR interface as in the Chap. 5 appendix,

```
import ogr
driver = ogr.GetDriverByName('ESRI Shapefile')
datasource = driver.CreateDataSource('towncenter.shp')
layer = datasource.CreateLayer('layerName',geom_type=ogr.wkbPolygon)
lonmin = -1.4366
```

```
latmin = 53.2242
lonmax = -1.4102
latmax = 53.2396
myRing = ogr.Geometry(type=ogr.wkbLinearRing)
myRing.AddPoint(lonmin, latmin)#LowerLeft
myRing.AddPoint(lonmin, latmax)#UpperLeft
myRing.AddPoint(lonmax, latmax)#UpperRight
myRing.AddPoint(lonmax, latmin)#Lower Right
myRing.AddPoint(lonmin,latmin)#close ring
myPoly = ogr.Geometry(type=ogr.wkbPolygon)
myPoly.AddGeometry(myRing)
feature = ogr.Feature( layer.GetLayerDefn() )
feature.SetGeometry(myPoly)
layer.CreateFeature(feature)
feature.Destroy()
datasource.Destroy()
```

Convert your shapefile to JSON format, used by Leaflet,

```
$ ogr2ogr -f GeoJSON -s_srs wgs84 -t_srs wgs84 \
    towncenter.json towncenter.shp
```

Next, manually add *"var towncenter="* to the start of the JSON file, and *";"* to the end. This provides a name for Leaflet to use.

Finally, create a web page which loads the Leaflet library, connects to an existing tile server, and loads and styles your JSON file,

```
<html>
<head>
  <link rel="stylesheet" \
    href="https://unpkg.com/leaflet@1.0.3/dist/leaflet.css" />
  <script src="https://unpkg.com/leaflet@1.0.3/dist/leaflet.js">
  </script>
  <style> #map { width: 600px; height: 400px; } </style>
</head>
<body>
<div id='map'></div>
<script src="towncenter.json" type="text/javascript"></script>
<script>

   var map = L.map('map').setView([53.2242, -1.4366], 13);
      L.tileLayer('<MYURL>',
   { maxZoom: 18, id: 'mapbox.light'
   }).addTo(map);

L.geoJSON([towncenter], {

   style: function (feature) {
      return feature.properties && feature.properties.style;
   },
   pointToLayer: function (feature, latlng) {
      return L.circleMarker(latlng, {
         radius: 8, fillColor: "#ff7800", color: "#000",
         weight: 1, opacity: 1, fillOpacity: 0.8
         });
      }
```

9.3 Exercises

```
    }).addTo(map);

</script>
</body>
</html>
```

You need an access token to make use of someone else's map tile server, e.g. you can obtain a free token from *www.mapbox.com* at the time of writing, having a form such as,

```
https://api.tiles.mapbox.com/v4/{id}/{z}/{x}/{y}.png?access_token=<MYTOKEN>,
```

and insert it into *<MYURL>*. You may need to pay if you make heavy use use of their server in a public system.

To view the result as in Fig. 9.23, open the *.html* file in a web browser.

9.3.2 Bluetooth Origin-Destination Flows

(This is a small project which links many areas of this book as well as visualization.)

Use the Derbyshire data to produce figures similar to the flow map of Fig. 9.16. Your program should match detected Bluetooth IDs at each possible origin and destination sensor, assign each such matched vehicle's journey to a route using Dijkstra routing as in Chap. 8, then sum the flows on each link and display them as thicknesses.

Hints:

A runnable version is provided in the software example, which uses 240 lines of Python code, which includes database setup, munging, matching, routing and display.

The task builds on, and reuses, code solutions from all of the exercises from previous chapters.

Most of the hard work is done by the database rather than Python. Some of the key parts of the supplied code include: recalling the origin and destinations of each route,

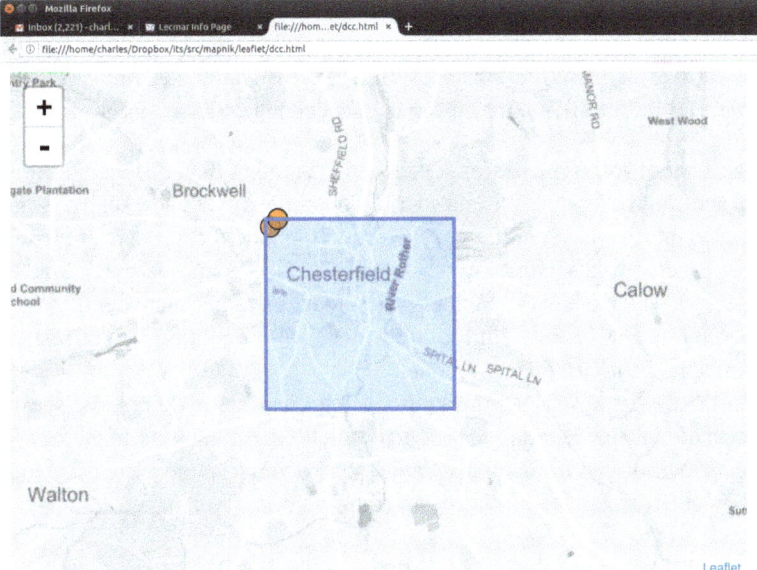

Fig. 9.23 Leaflet map

```sql
SELECT routeID,
  ST_X(orig.geom) AS ox, ST_Y(orig.geom) AS oy,
  ST_X(dest.geom) AS dx, ST_Y(dest.geom) AS dy
FROM Route,
  BluetoothSite AS orig,
  BluetoothSite AS dest
WHERE originSiteID=orig.siteID
  AND destSiteID=dest.siteID;
```

Matching Bluetooth detections at origin and destination to count the number of journeys on a route,

```sql
SELECT d.siteID AS dSiteID,
  d.mac as dmac,
  d.timestamp as dtimestamp ,
  o.siteID AS oSiteID,
  o.mac as omac, o.timestamp as otimestamp
FROM Detection AS d, Detection AS o
WHERE d.timestamp>o.timestamp
  AND o.mac=d.mac AND o.siteID='%s' AND d.siteID='%s';
```

Counting the number of journeys over each (link-broken) road link,

```sql
SELECT ways.gid, SUM(linkcount.count), ways.the_geom
FROM ways, linkcount
WHERE linkcount.gid::int=ways.gid
  AND linkcount.timestamp>'%s'
  AND linkcount.timestamp<'%s'
GROUP BY ways.gid;
```

Plotting road thicknesses by flow,

```
plot(xs, ys, color, linewidth=df_link['sum']/20000)
```

9.3.3 Large Project Suggestions

To make the sample code more realistic, as in Fig. 9.16, requires extra assumptions. Any or all of the following could be explored for MSc-sized projects, as undertaken on the ITS Leeds course:
Take the Derbyshire Bluetooth flow map concept and code and improve it. This could include:

- Take account of road type and/or congestion data rather than distance for travel times in routing.
- Detect and control for multiple and spurious counting of vehicles, such as when one vehicle appears on three or more sensors, or drives repeatedly back and forth past a single sensor.[5]
- Fuse the Bluetooth data with other sensor types included in the Derbyshire dataset, such as pneumatic road sensors which count only total flows at location rather than identifying unique vehicles.
- Replace shortest or fastest path routing with stochastic routing, or congestion (e.g. Wardrop equilibrium) approaches to route assignment.
- Use all the available sensor information, additional internet data, and sensible assumptions with Bayesian or machine learning models to optimally infer the state of the whole Derbyshire network over time and suggest traffic management approaches to improve it.
- Use Leaflet to display Derbyshire Bluetooth flows on a web page.

---

[5] As suggested by Lawrence Duncan to improve flow estimates.

For most transport applications, you can use Leaflet with an existing background map from a tile server like this, and overlay the visualizations that you want to present. Occasionally you might need to render your own cartographic quality map tiles yourself. This can be done using Mapnik – the same program used by OpenStreetMap to generate its own tiles, and by most other commercial tile servers. Details can be found at *mapnik.org*.

9.4 Further Reading

- Tufte E (1983) The visual display of quantitative information (Classic.)
- McCandless D (2012) Information is beautiful. Collins (So beautiful it is currently a mainstream best-seller)
- www.informationisbeautiful.net (Blog curated by the same author as the book.)
- Monmonier M (1996) How to lie with maps, 2nd edn. University of Chicago Press

Big Data

10

These is no standard definition of "big" or "small" data but we will define:

Small data sets are those which can be held and analyzed in a computer's memory, by consumer applications such as spreadsheets and scripting languages.

Medium-sized data is all the data held by a small or medium sized enterprise (SME), typically in one relational database in a single physical office. In 2018 this might include Terabytes of data. The SQL-based systems used so far in this book are of this level.

Enterprise data is when enterprise data outgrows a single database server, and uses ad-hoc scaling tricks to split or replicate relational data over multiple machines, covering systems in multiple offices or countries. In 2018 this might include tens of terabytes and millions of requests per second.

Big data is where parallel computing tools are needed to handle data. This represents a distinct and clearly defined change in the computer science used, via parallel programming theories, and losses of some of the guarantees and capabilities made by Codd's relational model. In 2018 this might include data centers containing tens of thousands of machines with petabytes or (for a handful of companies in the world), exabytes of data, as in Fig. 10.1. Most people who claim to work with "big data" do not actually do so in this sense (Fig. 10.2).

Boutique artisan small data is "real science" data collected causally by data scientists to complement and unlock larger found data sets.

10.1 Medium-Sized Data Speedups

If set up and optimized correctly, a single PC running a database server can often meet the needs for Terabyte scale data without the need for Enterprise or Big data level data systems. Think carefully before claiming that you need any more than one machine! Some options to consider here include:

- Use suitable data types in large and commonly accessed tables. Do you really need to use floating point numbers where smaller, faster integers can be used instead? Do you need large character arrays when smaller strings would suffice?
- Put indices on all columns that you need to search on to get from $O(n)$ to $O(\log n)$ search speeds.
- "Physical indices" are even faster than regular indices, only one per table can be applied, and ensures data is physically laid out on the storage in the index order for instant $O(1)$ access. They also do not take up additional storage space like regular indices do.

Fig. 10.1 Racks of parallel computers processing "big data"

Fig. 10.2 Key concepts in Big Data, according to #BigDataBorat

\	\	Data Science Conference Bingo Card	\	\
In-Memory	Unstructured Data	"We're Hiring"	Predictive Analytics	Streaming
Iris Data Set	Machine Learning	Real Time	Datafication	Facebook and Twitter
NoSQL	Mobile	Free Space!!	Internet of Things	Reuters-21578
Visualization	Hadoop	Social Graph	@BigDataBorat quote	Wordcount Demo
Sentiment Analysis	NCDC GSOD	Business Intelligence	Someone who thinks R doesn't suck	"Data Is The New Oil"

- Horizontal partitioning is where you split the rows of a table into multiple tables, to speed up searching for data if you have some feature of the search term that tells you which table to look in. For example, a *Drivers* table could be split into 26 tables called *DriversA, DriversB, DriversC*... where the letter is the first letter of their licence plate. If you need to look up a driver from the licence plate, you can tell which table to consult directly from its ID. If you are using $O(\log n)$ indices then you can reduce n this way.
- Vertical partitioning is another name for normalization, emphasizing the potential speedup when queries only relate to small aspects of entities which are captured in individual tables. (However, normalization can also slow the system down if you have to make lots of joins to recover everything you need.)
- Spatial data types (OGC shapes) will speed up searching for spatial data. You need to assign spatial indices, they are not automatic.

10.1 Medium-Sized Data Speedups

- Try to minimize the number of queries sent to the database as every query incurs some communications overhead. It is better to do heavy computing work using single, complex, SQL queries than working in your programming language with data from many queries. SQL query execution is highly optimized inside the database by very smart database programmers and will almost always go faster than anything you can write outside it.
- Complex SQL queries can themselves be massively optimized by considering in what order their parts and joins will be executed. Consider how your indices will relate to selections and joins, and what order these should run to minimize the work done by the database.
- Some database servers provide "in-memory caching" where commonly accessed data is moved from their hard discs to RAM for faster access.
- You can also write your client code to perform its own caching, either on the local hard disc or in memory, to reduce the number of queries and data sent over the network.
- While Codd's original relational theory prescribes using normalized data, with every fact represented only in one place, in practice your database will go a lot faster if you cache all data relevant to each analysis question into one big table. This breaks the nice theoretical consistency guarantees, but you don't care if you just want to get your computation to go faster.
- If you have really exhausted all of the above, then spend money on buying one high-end server computer to host the database. A RAID server has three hard discs storing copies of the data which can both speed up access and provide backup if one fails.

For example, the origin-destination analysis of London M25 ANPR data in Chap. 1 initially looked like a "big data" problem because it was taking the client's code hours to run single origin-destination queries. But rather than buying and installing new computers, we spent a few days tweaking the above speedups and got them to run in a fraction of a second. We could then could process a month of ANPR data over the whole M25 motorway in a few days on a single machine.

10.2 Enterprise Data Scaling

If you have really exhausted all of the above speedup options then the next step of scaling is to think about Enterprise level architectures. These typically split up different use-cases of the database into roles and data stores assigned to a small network (e.g. tens) of servers, such as illustrated in Fig. 10.3.[1]

Some of these roles might include:

Replication. If your data is fairly static (i.e. not being constantly updated from real-time sources) and is being used mostly for analysis, then a simple approach to handling many users is just to copy the whole database onto several independent servers. This is useful when you have many analyst users accessing data at the same time, but also if you can split your own analysis into small parts, run them on separate machines, then combine the results together at the end. For example, to compute the mean of a terabyte of floats, we could make 10 copies of the whole database, then ask each one to find the mean of a different 100 Gb subset of the data, then compute the mean of the means at the end. (This is a simple form of "map-reduce" which we look at later.)

Connection pooling. Rather than have users dial into specific replicated databases, a pooling server is a single public interface to the cluster of replicated servers which receives requests from multiple clients and automatically passes them to the best currently available replication server. Advanced poolers may consider and load-balance the computational capabilities of the individual servers, e.g.

---
[1] People who draw these diagrams have job titles like "Enterprise architect" and have some of the highest paid salaries in IT such as 100k+ GBP/year or 1000+GBP/day in 2018.

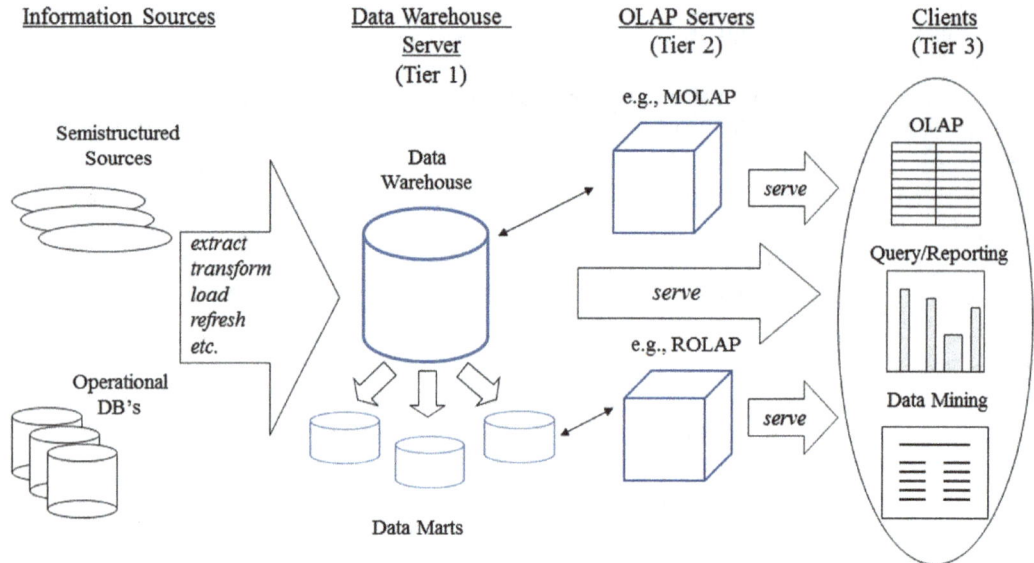

Fig. 10.3 Enterprise architecture

passing more queries to a server with a fast processor with lots of RAM than to an older machine in the same cluster. For postgres, the *pgpool* add-on does pooling.

ETL clients. Move your Extract,Transform,Load scripts onto different machines than the database server and have them do the text parsing themselves, and just send insert queries to the database server(s). You often want to schedule this to run overnight to minimize the server load for analysts working during the day.

If you need to handle inserts to the database, then replication is a problem because it may break the consistency of the database, as different copies get updated at different and often unpredictable times, because they are busy doing other work serving users. In these cases, more advanced options are:

OLAP architecture. As illustrated in Fig. 10.3, OLAP architectures use separate machines for the main data storage ("warehouse"), for real-time operational data (such as live dashboards) and for analysis (on-line analytic processing, OLAP). OLAP tables are denormalized, data caching tables as in the single-server case, but different formats optimized for different queries or classes of queries can be hosted on separate machines. Sometimes they will be completely unnormalized for speed (multidimensional OLAP, MOLAP) at the expense of losing flexibility; other times some relational structure will be preserved for flexibility at the loss of speed (relational OLAP, ROLAP). Jobs that create OLAP tables from warehouse data should run overnight to avoid load during analysts' work.

Tier3 OLAP. If OLAP gets really big, you can create OLAP tables from other OLAP tables, on new machines. For example if a group of analysts want to work with related data but each in their own ways.

Data marts are subsets of the main data warehouse which are typically accessed together, and can be usefully stored on databases in separate machines. For example, all traffic sensor data on one server, all car insurance policy data on another. Whilst dividing the labour for common tasks, this makes it difficult to combine data between two marts when they need to interact. If such stores are not able to talk to each other very well, the derogatory term "data silo" is sometimes used.[2]

---

[2] Typically by consultants who want to charge you a lot of money to do something about them. Often the same people who draw enterprise diagrams.

10.2 Enterprise Data Scaling

Operational data where live inserts and real-time access is needed, such as for live dashboards, can be kept on dedicated servers, with overnight scripts moving it into the warehouse every night or at other quiet times.

Sharding is horizontal partitioning with the tables stored on separate servers. Tools such as *pgshard* can help to automate this. In some cases, sharding can be made to appear transparent at the ontological (table design) layer, being handled by the physical database implementation, whereas basic partitioning requires an ontological split into tables of different names.

Networking hardware can speedup up connections between machines in the cluster, for example using fast Infiniband instead of Ethernet links. This can get very complicated, especially if the enterprise has sites in different physical locations which need to access the same data and are connected over the Internet at slower speeds than links within clusters. Specialist file systems like Lustre can distribute and load-balance file storage over multiple hard discs like RAID, but over physical file server machines in a cluster.

10.3 CAP Theorem

Enterprise architectures developed in an ad-hoc way, with relatively little understanding of their computer science properties. The original Codd model is very prescriptive – it takes a "modernist" view of data as a clean, consistent picture of the world. All of the Enterprise methods violate at least one of Codd's principles, and can result in cases where inconsistent data can appear in the database.

The Codd model arose at a time when there was less distinction between database research and what is (properly) called Artificial Intelligence, AI. The goal of AI was to represent the world of facts as logic, which could be used to answer very complex logical queries, typically using logic formalisms such as Prolog. Prolog programs are lists of logical facts and deduction rules, such as the following model of a car accident and rule to determine if any insurance companies need to call each other,

```
person(ludvig).
person(edgar).
car(car1).
car(car2).
driver(ludvig, car1).
driver(edgar,car2).
crash(car1, car2).
insurer(ludvig, bigactuaryco).
insurer(edgar, bigdataco).
tocall(A, B)   :- driver( X, XC) , driver( Y, YC) , \
          crash(XC, YC), insurer(X,A) , insurer(Y,B)  .
```

(The symbol ":-" means roughly, "is true, if", and the comma "," reads as "and".) This logical, Wittgensteinian, view of AI placed emphasis on logical deduction algorithms, rather than on data storage or efficiency, though is (roughly) equivalent in power to what SQL can do. But relational SQL databases emphasize storage and efficiency rather than logic in their implementations. Prolog-style AI went out of fashion for a long time but SQL survived – doing similar things but in a more practical and less hyped way.

But the legacy of AI on relational theory has been the continued emphasis on data consistency. From an AI view, a single contradiction stored in the database could logically prove any statement such as *True* = *False*, as in mathematics.[3] In modern data analysis there is less emphasis on logic or consistency – everything is noisy, continuous-valued, and probabilistic and it usually doesn't matter if there are a few data glitches.

This provides some understanding of what has been happening in the real, industrial world where Enterprise architectures have evolved. The ad-hoc extensions are messy and destroy Codd's desired consistency properties – but industry generally doesn't care because it is no longer using logical AI inference methods that rely on consistency. Companies have been willing to trade off the nice theoretical properties in order to build systems that work and serve their users at scale in practice. Until recently, this was embarrassing for Theory, which mostly ignored the real world and continued to teach Codd's view.

Recently though, Theory has analyzed these deviations from Codd's model in detail. The CAP theorem (Gilbert and Lynch 2002) has proved that it is not possible to build databases spanning multiple servers that have the three properties of Consistency, Availability, and Partition-tolerance. Availability means that any query receives some answer, and Partition-tolerance means that any network link may go off-line.

10.4 Big Data Scaling

Following the CAP theorem, much research has questioned Codd's requirements, and designed new databases that trade off one or more of the CAP properties in exchange for the ability to scale up across multiple servers. This is an exciting time for database design, with new ideas and software now being proposed all the time, after almost 40 years of the Codd model. Unlike ad-hoc Enterprise architectures, which rely on the use of small number of machines, these big data systems may be almost infinitely scalable over servers, because they are free of the Codd assumptions that required data stored in different places to be join-able together.

10.4.1 Data "Lakes"

One option is to abandon the idea of a database altogether, and go back to storing raw files as discussed at the start of Codd's paper. By storing raw, unprocessed input data across multiple servers as a "data lake", we allow the data analyst to perform their own ETL type operations on the fly, rather than forcing any particular ontological interpretation of the data onto them. If the Wittgenstein-Codd view of the world as carefully curated, consistent facts is "modernist" then we could say this is a "post-modern" data architecture, after those philosophers such as Gadamer who have made claims such as,

> "meaning for us emerges from a text only as we engage in a dialogue with it."

This approach also allows the use of unstructured data such as images and natural language to be analyzed directly by expert analysts without the need for difficult ETL by their DBA.

---

[3]This view of AI is the source of the old media caricature of robots crashing with "does not compute" error when faced with contradictions. Later AI research did manage to avoid such spectacular logical blow-ups by using different logical foundations such as para-consistent logic in "Truth Maintenance Systems". In these systems, you might prove that the car is red and also prove that the car is blue, but it no longer follows from this that $0 = 1$ because the contradictions are contained in logical domains, and indicate problems with your assumptions rather than the state of the world. This is based on human reasoning, which very often can deduce contradictory results, but also without them exploding. See (Doyle 1979, *A truth maintenance system*, Artificial Intelligence 12(3):231–271).

Some disadvantages of this approach are: speed of computation – running ETL on the fly for every query is much slower than just doing it once when data is imported by a DBA; and speed of software development, if analysts need to write their own ETL scripts and manage SQL-like operations themselves. Some classical database administrators and architects may have difficulty with empowering their analysts to work with raw data in this way. As a culture, they have grown used to being powerful employees who define highly centralized systems and ontologies that shape the processes and thinking of their organizations. According to "The McKinsey Mind", this kind of conceptualizing and problem-defining is a key part of senior management, and not something that would be left to analysts.

10.4.2 Grid Computing

Scientific computing has operated on lake-like data for several decades in systems known as grids. For example, speech recognition research never switched from flat files to databases at all, and has always stored large numbers of raw audio and text items as individual files. Often in scientific computing, the data consists of large numbers of similar entities, which do not relate to each other very much or at all. Again in speech recognition, each audio-text pair of speech and transcript can be processed completely independently of the others, in sequence. There is no selecting or joining of data. Such computation tasks are called "trivially parallelizable" because they can can very easily be split into independent jobs running across a cluster of identical machines. Each machine is given a subset of the data to work on, and returns its results to a master server for collation. For example to compute the average error of a speech recognizer over 1Tb of data, we divide it into 1Gb sets and give ("map") each set to one of 1,000 machines. Each machine $i \in (0, M)$ returns its own average error $\bar{x}_i$ and the number N_i of cases tested on it. The master server then "reduces" these statistics to a single mean via,

$$\bar{x} = \frac{\sum_{i=0}^{M} N_i \bar{x}_i}{\sum_{i=0}^{M} N_i}.$$

- How would you compute the standard deviation in a similar way?

The process of making the data partitions, assigning them to servers, load balancing and scheduling work requests from many users, and collating their results may be done manually by your own programs (the simplest form is just to split your data into 10 folders, copy each folder onto a laptop, and run your script on each one manually), or more usually using a "Grid Engine" such as SGE (Sun Grid Engine) to automate parts of it. SGE allows you to interact just with one master node, request work to be done, and have it (almost) automatically split up, sent off to servers, and collated.

There are three well-known problems with Grid Computing. First, the data usually lives on a single file server machine, or sometimes on a few clustered file servers using tools such as the "Lustre" file system. Data does not live on the same machines where the computations take place. So whenever a computation is run, large chunks of the data must be transferred across the cluster network to each of the computing nodes. Even if fast networks such as Infiniband are used, this will not scale part a certain size when the network becomes congested. Network bandwidth is the bottleneck.

Second, grid engines such as SGE are not fault-tolerant. If a server goes down during a large computation (which can be very common in a busy cluster), it may invalidate the entire computation including the work of other nodes; or in the best case will require manual intervention to set up and re-run the missing parts again.

Third, grid engines are designed to run on well-defined, physical clusters of servers, such as a room in a university full of servers. Once you have chosen the size of this cluster (e.g. 1000 servers) and spent time setting up the grid engine, it is not easy to add new capacity on-the-fly, other than my buying new physical servers, hiring someone to install them, and usually taking the grid off-line for a day during that installation process. The lack of scaling is a particular problem when the grid is shared by many users and groups. Famously in universities, a computation heavy group such as CERN Higgs boson searchers will often completely take over a grid for weeks at a time around their conference seasons, rendering it useless to everyone else, unless social engineering (such as cash transfers) is done to balance the demand.[4]

At many large institutions such as universities, SGE is available on thousands of nodes is available through managed, centralized IT services. In some cases you may need to demonstrate competence before being allowed to use them, because it may possible to make interactions between different people's jobs or otherwise break things for other users. You can use SGE with Python by writing scripts like this which read their task IDs from SGE, and use them to read, process, and write different files to the shared file-system. (You would then write a reducer script to collate all the output results too),

```
#myJob.py:
my_job_ID = os.environ['SGE_TASK_ID']
input_data_filename = '/home/charles/data_%i.csv'%my_job_ID
output_results_filename = '/home/charles/results_%i.csv'%my_job_ID
myProcessingFunction(input_data_filename, output_results_filename)
```

then uses the *qsub* command to launch, say, 1000 of these jobs in parallel across the cluster,

```
$ qsub -t 0-1000 myJob.py
```

10.4.3 Map-Reduce and Cloud Computing

The above three problems with grid computing are fixed by the Map-Reduce frameworks. These are software tools which enable you to write similar "map" and "reduce" scripts as used in grids, but process them in more advanced ways.

First, they distribute the storage of the data across all servers in the network, rather than it living on a small number of file-servers. They arrange the computation load so that computation on data is performed on the same servers that host it, so no data needs to be transferred over scarce network bandwidth during the map phase. This removes the network bottleneck and also removes the need for expensive fast networks such as Infiniband.

Second, they are fault-tolerant. The distributed data storage includes copies of each data segment on multiple servers. (The concepts of distributed copies is similar to Lustre, but here the servers are being used as computation nodes rather than just file servers as in Lustre.) If data storage or computation fails on any server, the Map Reduce framework detects this, makes new copies on other nodes, and automatically re-runs the computations there. Fault-tolerance not only allows for failures of high-

---

[4]This makes life hard for other researchers who happen to have their own conference seasons at the same time, and it is not unknown for dirty tricks to be played to optimize one group's compute time at the expense of other's when this happens. For some reason, physicists always seem to get top priority. Depending on your point of view, this may be because their work is more fundamental to the progress of human knowledge, or because they got the funding to set up the cluster.

performance computers, but can remove the need for them completely and allow computation to run on large clusters of cheap consumer grade compute hardware.

Third, they are designed to be easily and infinitely scalable. Typically you run a Map-Reduce clusters on some company's servers in their data center (a "cloud"), renting them per hour rather than maintaining your own hardware. These centers may have millions of computers which are used for many purposes by many people, but by economies of scale (and/or by paying them more money) you are almost always able to buy time on additional machines without worrying about what other users are doing. Because the only data passed around during normal operation is the mapper outputs, rather than the big data itself, this does not put such a strain on the network bandwidth or act as a bottleneck.

These design concepts, when taken together, give rise to a more "anarchic" or "open-source" style of cloud computing than found in grid computing. Grids must be centrally planned and maintained, using expensive high-performance hardware and dedicated IT professionals to keep them running. Map-reduce in contrast runs on cheap consumer hardware and is designed to allow and work around failures. It is non-trivial to add extra capacity to a grid but it is simple to add new computers to a map-reduce setup. Usually to work with a grid you must be a member of a large organization which has a central grid and full-time managers, but anyone can set up a small, cheap map-reduce system themselves. Map-reduce's ability to quickly reconfigure to work across clusters of added and removed machines make it a good fit to cloud computing, in which one to millions of cheap computers are rented per unit per hour from external companies by users, rather then set up in house. This style of processing has replaced grids for many Data Science tasks for these reasons. However grids still have strengths over clouds for other task types, such as solving problems which are computationally intensive rather than data-intensive, and tasks which do not easily decompose into map-reducible units.

10.4.4 Hadoop Ecosystem

The best known map-reduce systems are Hadoop and Spark, which perform map-reduce and related computations. Hadoop was the original map-reduce component of what is now a large ecosystem of open-source big data software. Modern implementations run on top of the tools HDFS and YARN. HDFS (Hadoop Distributed File System) appears to the user as a single big hard disc where very big data files can be stored, but internally breaks such files into small pieces and distributes their storage and computation across the cluster (roughly analogous to Lustre in grid computing). YARN (Yet Another Resource Negotiator) provides interfaces to request computation resources across the cluster (roughly analogous to SGE in grid computing).

A recent trend has been to move data off discs altogether and work "in-memory" in server RAM, which is especially useful for machine learning applications requiring many passes over the same data, for example during parameter optimization. This is implemented by the Spark system.

Other recent components of this ecosystem include tools to transform many common data tasks into map-reduce forms automatically; tools to move data in and out of HDFS, perhaps in real-time (Flume); tools for automating complex work-flows comprised of many map-reduce jobs (Cascading, Tez); and more advanced interfaces to various languages (*mrjob*, *pydoop*). Other parallel computation architectures than map-reduce have been implemented on YARN, for example, the DL4J library builds parallel distributed Hierarchical Linear In Parameters Regression ("deep learning"/"neural network") units on YARN, and we may see large Bayesian Networks become implemented similarly.[5] Other tools, such as the Impala database, are being developed based only on HDFS without either map-reduce or

---

[5]*blog.cloudera.com/blog/2014/08/bayesian-machine-learning-on-apache-spark/.*

YARN. This "Hadoop ecosystem" changes very quickly – despite retaining its name many systems no longer use the original Hadoop software at all, building instead on HDFS and/or YARN – as it is constantly developed by many of the big Silicon Valley companies. See *hadoop.apache.org* for the latest developments.

10.4.5 Non-relational Databases ("NoSQL")

While grid and map-reduce systems can work with general unstructured "data lakes", in practice they are almost always used to operate on lists of similar entities. For example, Hadoop file rows will usually all have the same format, such as the same set of CSV-like fields. This is because as databases scale up they usually contain "more or the same" – more data collected in the same ways about the same things in the world, rather than new types of facts about new things. Most of the world's biggest databases are like this, having rows of customer transactions, car journeys, or government citizen records.

The way that such data is usually accessed is by assuming that each row models one entity, with properties. Ontologically, we saw earlier in this book that there has been ongoing tension between the Aristotelian, object-oriented, view of the world as made of entities and properties; and the Wittgensteinian, relational view of the world as made of logical facts. The relational view is more powerful, because it allows relations to be combined in arbitrary ways to form new relations, while the entity-property view is more rigid, fixing the ontology in advance. A practical move towards data tables representing distinct entities rather than more general relations has been going on for some time. Programmers have not found the general relational model to be as practically useful as Codd and AI researchers had intended. Logic and logical AI have generally fallen out of fashion, having been replaced by probabilistic and classification interests, so the ability to combine logical facts in complex ways is less in demand. Financially, most data interest now comes from commercial web companies wanting to process logs of shopping baskets and clicks rather than to understand intelligence. And perhaps the popularity of object-orientation in programming languages (as distinct from databases) has been involved too. If we assume one class per table then it is simple to store and retrieve software objects in persistent databases. You have probably found yourself writing lots of code like this by now,

```
sql = 'SELECT * FROM sites WHERE ... ;'
site.easting = df['easting']
site.northing = df['northing']
site.installDate = df['installDate']
```

which is a painful way to convert between relational databases and object-oriented programming. However, if we assume one class per table, then it is possible to automate such conversions using Object Relation Mapper (ORM) libraries such as SQLAlchemy. Other Object-Oriented Databases (OOD) are built from the ground up specifically to support this style of programming. This only works if we ban on-the-fly relation combination, making the distinction between "real" entities which have their own tables, and "second class" relationships between entities which are not explicitly represented by tables and classes.[6]

Giving up relational combination makes it much easier to design distributed databases, because most of the complexity of a distributed relational system comes from having servers communicate to join data stored on separate machines. If all we need from a database is the ability to store and retrieve

---

[6]This may depend on the style of object orientation used by the programming language too. In strongly-typed languages like C++ all classes used in the program must be predefined, but a "duck-typed" language like Python is able to construct new classes on the fly at run time which could in theory be used to represent arbitrary relations as objects. I don't know if anyone has tried this yet?

specific entities, rather than relations, then we can simply shard both horizontally and vertically, and each query can be executed on the relevant servers independently of one another. This is "trivial parallelization" as in grids and map-reduce. The CAP theorem says that we risk losing consistency at times by doing this, but if you are working will terabytes of historical customer shopping baskets you don't usually care if a few of them are a little out of date or contradict one another, in the way that an logical AI researcher would care.

Some database designers have built systems under this assumption, but have tried to retain parts of SQL syntax to make them easily usable – as most database users are trained in SQL. These systems typically appear as subsets of SQL, providing features like *SELECT*, *WHERE*, *COUNT* and *GROUP BY* but omitting *JOIN*s. They have come to be known as "NoSQL" systems.

NoSQL systems have been built to optimize many different functions at the expense of others. Some make access to historical data very fast by not caring about real-time insert updates, other optimize for operational use, and others for data-mining type queries which can be assumed to request the same data subsets many times. Some are optimized for reading, or writing, or for balancing both reads and writes. Some abandon indexing or selecting by property values to focus on simply retrieving entities from their names (primary keys). Some "value-stores" enforce the class-table mapping by forcing all rows to have the same structure (Redis, Cassandra[7]); other "document store" databases are looser, lake-like models which simply map indexed names to arbitrary, size-unlimited collections of zeros and ones or JSON structures (MongoDB) to be used in any way the programmer chooses.

10.4.6 Distributed Relational Databases ("NewSQL")

After many decades of dominance by the Codd model and SQL, the last ten years have seen an explosion of new database designs which is still ongoing. New databases and versions are being released every month, and as most are open source, they are also forked into thousands of experimental versions every day. It is hard or impossible to keep up with them all, and anything written about specific systems is likely to change before it is read! Older databases like Postgres do not remain static either, and their designers watch the NoSQL systems carefully and borrow their ideas and features, so they can longer be categorized as purely relational themselves. Integration of spatial and GIS SQL extensions will need to keep up with the new systems.[8]

A current research area aims to develop versions of full Bayesian inference systems over big data sets, such as Bayesian networks whose nodes are implemented on different machines communicating by passing messages over the network.[9]

Another research area tries to reconstruct as much of relational theory as possible in distributed systems. We know that it is impossible to violate the CAP theorem, but by relaxing some of its assumptions, such as allowing for brief inconsistencies or outages, much of the classical model can be put back together. Apache Hive is a database that runs on top of Hadoop, using an SQL dialect, HiveQL, and able to perform joins. Cloudera Impala is another database which implements the same HiveQL, running faster without Hadoop, directly on its underlying HDFS. The computer science of this is very, very hard, requiring optimization of both network and computational resources to move bulk data around nodes during query processing. This may remain an active research topic for a long time.

---

[7] Also LMBD the "Lightning Memory Mapped Database" has this form, though optimized for very fast retrieval on a single machine – useful for machine learning training.

[8] For example, Hadoop *esri.github.io/gis-tools-for-hadoop*.

[9] For example, *http://blog.cloudera.com/blog/2014/08/bayesian-machine-learning-on-apache-spark/* is a project to run distributed PyMC on Spark.

But if and when the reconstruction is finished, data analysts will be able to go back to using classical SQL exactly as in the 1970s, without caring what new computer science algorithms are involved in answering their queries.

10.5 Exercises

10.5.1 Prolog AI Car Insurance Queries

You can run the above Prolog program of sect. 10.3 as follows,

```
$ gprolog
[user].
<type the program here>
<press Ctrl-D>
tocall(X,Y).
```

The final command asks the "knowledge base" consisting of the logical facts provided, if it can find any values X *and* Y for to make the relation *tocall(X,Y)* true. (Note that Prolog commands end with a full stop.) It should find the names of all pairs of insurance companies which need to call each other due to crashes of vehicles driven by their insurees. Extend this model to have more facts and queries. How is Prolog similar to SQL and how is it different?

10.5.2 Map-Reduce on Vehicle Bluetooth Data

The best-known map-reduce framework is currently Hadoop, together with its distributed file system HDFS. In a standard setup, all the data is represented as one huge file (e.g. many terabytes or petabytes). Inside HDFS, the data from this file is split up and stored across the servers, but it appears as a single file to the analyst. Hadoop assumes all files are text files, and contain one data item per row, as in CSV. Input rows can have any format, but Hadoop's own files usually use "Key-Tab-Value" rows. This means a text ID (or "key") comes first, then a tab character, then arbitrary value data. It is possible to but anything in the value data, including binary data such as images and videos, as well as CSV-like fields.[10]

Suppose we have a big data (e.g. several Tb) set of Bluetooth detections, which arrive as CSV files containing detection DateTime and Bluetooth MAC identifier of the detected devices as in *data/dcc/bluetooth/*.csv*,

```
14/02/2017 00:03:08,MAC,,90,196 (MAC),C292952806B7
14/02/2017 00:04:18,MAC,,90,196 (MAC),1174744003D2
14/02/2017 00:04:18,MAC,Reverse,90,196 (MAC),1174744003D2
14/02/2017 00:04:19,MAC,,90,196 (MAC),F89490400987
14/02/2017 00:04:41,MAC,Reverse,90,196 (MAC),C292952806B7
14/02/2017 00:04:44,MAC,,90,196 (MAC),8054554004F8
...
```

---

[10]However, Hadoop is critically dependent on the use of tab, newline and carriage return characters. Binary data may contain these in some bytes, which you need to remove. One way to do this is to replace them with some other long strings using regexes, then replace those strings with the original characters when you are ready to process the data in your mapper. Alternatively, libraries and "SeqFiles" can be used to do something similar.

We would like to find the total number of times a particular Bluetooth MAC has been detected. We can write and test mapper and reducer functions on a local PC using a short extract from this data, without needing an actual Hadoop cluster as follows. Here is a *mapper.py*,

```python
#!/usr/bin/env python
import sys
for line in sys.stdin: #read from input stream
   line = line.strip()
   (timestamp,type,heading,foo,bar,mac) = line.split(",")
   count=0
   if mac=="F89490400987":     #the MAC we seek
      count=1
   print('%s\t%s' % (mac, count))    #output KTV
```

Test the mapper by streaming a single row string to it, then by sending a whole *.csv* file to it,

```
$ chmod +x mapper.py
$ echo "14/02/2017 00:00:13,MAC,Reverse,89,194 (MAC),
F89490400987" \
   | ./mapper.py
$ cat /data/dcc/bluetooth/vdFeb14_MAC000010100.csv | ./mapper.py
```

(The "#!" code line and the *chmod* command make the program usable by Hadoop and other tools.) This outputs KTV data such as below (where the space between the ID and the value is a single tab character). The mapper has appended a '1' to each instance of the desired MAC:

```
EAB3B138083C 0
F89490400987    1
019096000736 0
...
#!/usr/bin/env python
from operator import itemgetter
```

Write a *reducer.py* which sums the mapper outputs for each MAC,

```python
import sys
current_mac = None
current_count = 0
word = None
for line in sys.stdin:

   line = line.strip()
   mac, count = line.split('\t', 1)
   count = int(count)
   if current_mac == mac:
      current_count += count #as inputs are sorted
   else:
      if current_mac: #output in KTV format
         if current_count>0:

            print('%s\t%s' % (current_mac, current_count))
      current_count = count; current_mac = mac
```

Test the reducer on sorted mapper output,

```
$ cat /data/dcc/bluetooth/vdFeb14_MAC000010100.csv | \
  ./mapper.py | sort -k1,1 | ./reducer.py
```

To give output such as,

```
F89490400987      170
```

10.5.3 Setting up Hadoop and Spark

Installing physical Hadoop and Spark clusters is beyond the scope of this book and beyond most data analysts. It is usually done by a dedicated IT specialist. It is is hard, because these tools are designed to be manually configured to run on thousands of computers, possibly in different rooms or countries, and with different reliabilities and speeds, and with strong security requirements.

Rather than require the reader to access a compute cluster or perform such an installation, a second Docker image is available from Docker's (or this book's) server which simulates a Hadoop and Spark cluster on a single machine. It is called *itsleeds/itsleeds-bigdata* and can be downloaded and run from Docker similarly to the main *itsleeds* image via,[11]

```
docker run -it -p 8088:8088 -p 8042:8042 \
  -h sandbox itsleeds/itsleeds-bigdata  bash
```

Unlike *itsleeds*, but in common with most real compute clusters, there is no graphical interface and all interaction must be done through the command line. The extra arguments in the command are required because big data tools make use of advanced networking features of physical and virtual machines. If you wish to run on a real Hadoop/Spark cluster, commercial services such as Rackspace (*http://go.rackspace.com/data-baremetalbigdata.html*) and Amazon (*https://aws.amazon.com/emr/details/spark/*) provide access to pre-configured systems for a fee, at the time of writing.

10.5.4 Finding Vehicle Matches in Hadoop

Once set up, the same mapper and reducer as tested above can be run on a cluster as follows. First, we copy the data to the HDFS file system,

```
$ hadoop fs -copyFromLocal  \
  data/dcc/bluetooth/vdFeb14_MAC000010100.csv \
  vdFeb14_MAC000010100.csv
```

Tell Hadoop to run the map-reduce job (via its "Streaming" mode which connects up the input and output string streams),

```
$ hadoop jar $HADOOP_HOME/share/hadoop/tools/lib/hadoop-streaming-2.7.3.jar \
  -file mapper.py -mapper mapper.py -file reducer.py -reducer reducer.py \
  -input   vdFeb14_MAC000010100.csv -output myout.ktv
```

---

[11] Native Linux users should prefix with *sudo*.

10.5 Exercises

Wait for the job to finish, then recover the output file from HDFS to see the results,

```
$ hadoop fs -get myout.ktv myout.ktv
$ less myout.ktv
```

This will give the same result as in the previous example – but this system could be used to search for the detections over petabytes of data.

Try changing the mapper to output a "1" for *every* Bluetooth MAC – feeding this to the reducer will then compute counts of sightings of all vehicles on the network on this day.

Try counting every Bluetooth device detected at every survey site in Derbyshire on the survey day, with Hadoop. You can merge the contents of multiple files into a single file with commands like,

```
cat data/dcc/bluetooth/vdFeb14_MAC000010101.csv \
    data/dcc/bluetooth/vdFeb14_MAC000010102.csv | \
    data/dcc/bluetooth/vdFeb14_MAC000010103.csv | \
    grep -v "Number Plate" > all.csv
```

The *grep* part of this command removes the header lines (which contain the phrase "Number Plate") from each .csv file during the merge. The output is stored in the new file *all.csv*.

10.5.5 Traffic Flow Prediction with Spark

Here we will try to predict the traffic flow at one location given readings over time from a set of other locations and no other information about the network. This is an example of a purely data-driven model, which makes no reference to spatial concepts such as maps, routing or Wardrop equilibria.

Suppose we have observed Bluetooth traffic flows $b_{i,t}$ at locations $i = 1 : N$, in one-minute time bins t. We have $N = 7$ sensors around Chesterfield. Suppose sensor $i = 0$ has broken down and we wish to estimate the flows there, $\hat{b}_{0,t}$. We model these estimates as a linear function of the other sensors' data,[12]

$$\hat{b}_{0,t} = \sum_{i=1}^{N} \sum_{\tau=0}^{T} w_{i,\tau} b_{i,t-\tau}.$$

The theory here is that the flow at site 0 includes some portion of traffic from each of the other sites from some time in the past, because it has taken time to travel from there to site 0. Suppose that the maximum travel time between site 0 and any other site is $T = 5$ (min). We wish to find the parameters $\{w_{i,\tau}\}_{i,\tau}$. There are $(N - 1)(T + 1) = 36$ parameters.

The model is linear and can be solved by linear regression. In theory, linear regression is trivial. For small data you can just run a standard regression program like the one in the Python *sklearn* library. (Or in Matlab, simply type "$b = y\backslash X$" !) In practice there is lots of data which must be (a) preprocessed and (b) computed, e.g. using distributed parallel software like Spark. It might take hours or years to run in *sklearn* on a single computer.

---

[12] As suggested by ITS Leeds student Aseem Awad. More generally and usefully this could be used to predict total flows as counted by sensors such as temporary induction loops, after the temporary sensors have been used for calibration then removed.

Spark has a machine learning library which includes a linear regression function, able to run at scale (e.g. on petabytes of observation data). Like most Spark programs, this function takes (conceptually) a simple text file as input. Each line of the text file has the format like this,

```
26.0 1:77 2:77 3:0.0 4:0.0 5:0.0 6:0.0 7:42 8:42 9:0.0 10:0.0 \
11:0.0 12:0.0 13:185 14:185 15:0.0 16:0.0 17:0.0 18:0.0 19:63 \
20:63 21:0.0 22:0.0 23:0.0 24:0.0 25:22 26:22 27:0.0 28:0.0 \
29:0.0 30:0.0 31:32 32:32 33:0.0 34:0.0 35:0.0 36:0.0
```

The first number is the prediction target, $\hat{b}_{0,t}$ (with rows of the text file ranging over t). The other 36 numbers are the $(N-1)(T+1)$ regression inputs, $b_{i,t-\tau}$. (This format is known as *libsvm* format.)

There may be zillions of rows of this text file from many observations. In practice it is implemented as a complex structure known as an RDD rather than an actual text file. For small examples such as here we can use an actual text file though. Inside Spark (and the many other libraries that it builds upon, including Hadoop's HDFS), the rows of the RDD are split up into chunks of rows, and distributed (and replicated, in case machines crash) between many (e.g. thousands) of machines across a compute cluster. Most machines then runs subproblems of the linear regression task, using their own chunks ("map"), while a few other machines read the results from those chunks and combine them ("reduce") into the final solution. In our toy example we will run on a single machine, but using the full Spark setup. We would use exactly the same Spark commands to run on a full cluster if we had more data.

Tasks:

The first task is to prepare the input text file for Spark, *data.libsvm,* containing rows in the above format, using the Derbyshire Bluetooth data. To do this requires us to look up flows from previous times to the time being predicted. One way to do this is to read all the flows into a database, then use SQL queries to retrieve flows at the required times to create each line. (If you prefer not to do this and use Spark straight away, a pre-built *data.libsvm* is provided in *codeExamples.*) Upload it to HDFS with,

```
$ hadoop fs -put data.libsvm
```

To use Spark, type,

```
$ pyspark
```

This may take a few minutes to run (because Spark is checking whether there are thousands of other computers connected to it and how to command them). When this competes you should see an ASCII Spark logo and a Python command line as shown below. You can import the linear regression library and run regression on your data file as shown in the commands below. The resulting coefficients, $w_{i,\tau}$ are printed out at the end.

```
Welcome to       ____              __
     / __/__  ___ _____/ /__
    _\ \/ _ \/ _ `/ __/  '_/
   /__ / .__/\_,_/_/ /_/\_\   version 2.2.0
      /_/
Using Python version 2.7.12 (default, Nov 19 2016 06:48:10)
SparkSession available as 'spark'.
```

To load your data from HDFS, type,

```
training = spark.read.format("libsvm").load("data.libsvm")
```

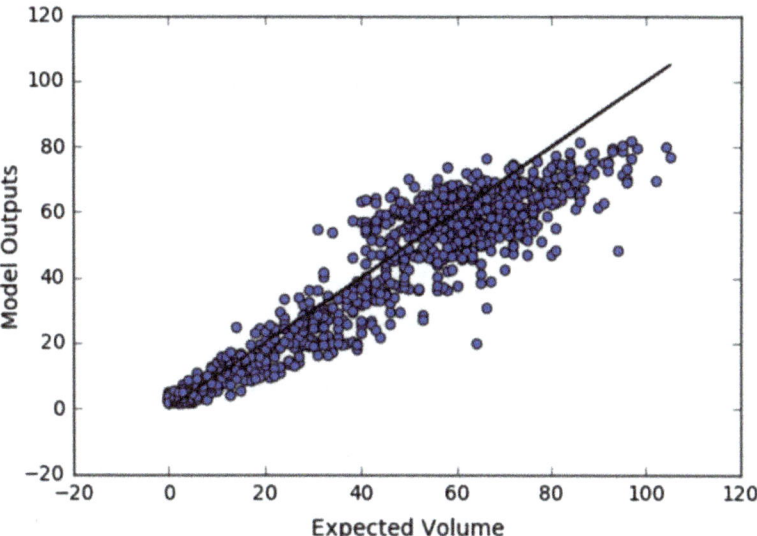

Fig. 10.4 Spark regression model output, showing predicted flows at all Bluetooth sites against corresponding ground truth flows. Model and figure by Leeds MSc student Aseem Awad

Then to define and fit the model,

```
from pyspark.ml.regression
import LinearRegression
lr = LinearRegression(maxIter=10, regParam=0.3,
elasticNetParam=0.8)
lrModel = lr.fit(training)
# Print the coefficients and intercept for regression
print("Coefficients: " + str(lrModel.coefficients))
print("Intercept: " + str(lrModel.intercept))
```

This will show the resulting parameters. To show an R^2 statistic (i.e. the percentage of variance explained by the model) for the linear regression,

```
trainingSummary = lrModel.summary
print("r2: %f" % trainingSummary.r2)
```

Note that the (classical, non-Bayesian) R^2 statistic should be adjusted to account for over-fitting when large numbers of parameters: "adjusted R^2" is given by,

$$R_a^2 = 1 - \frac{n-1}{n-k-1}(1-R^2), \qquad (10.1)$$

where n is the number of data points (rows of the *libsvm* file) and k is the number of parameters (columns of the libsvm file minus one). This represents a trade-off between penalizing the ability of large parameters (k) to over-fit the data, and rewarding the model's ability to explain lots of data (n), similarly to the BIC in Bayesian statistics.

The regression can also be plotted to show the fit as in Fig. 10.4.

The *spark.ml.regression* and *spark.ml.classification* libraries also have many other discriminative (aka. "machine learning") functions, including nonlinear regression methods like decision trees and neural networks, which all have similar interfaces to the linear regression function. (Classification is just a special case of regression with targets taking Boolean or integer values). Try swapping them in to replace the linear regression and see which models give the best predictions of the data. See the documentation at *https://spark.apache.org/docs/2.1.0/ml-classification-regression.html*. If you are able to access a larger Spark cluster, try running the same code on a larger data set with larger (e.g. terabytes of data) values of N, T and t.

10.5.6 Large Project Suggestions

Obtain additional Bluetooth data for many days and use Bayesian or machine learning models to classify individual Bluetooth-tracked vehicles into categories such as (1) daily commuters (who almost always take the same route at the same times of day); (2) tourists (who only visit Derbyshire for one day); (3) irregular local road users (who appear many times on the network but taking different routes). Plot flows for these market segments. Make recommendations to improve network usage based on them.

Build a Markov Random Field model of the network based on link-broken vertices as nodes and roads as connections, and use an EM algorithm to infer both flows at all vertices and correlations between vertices in the road network. How can Dijkstra-like routing priors be fused into such a model?

Consider non-physical interventions to the network, for example can flows be improved if public-sector commuters' shift patterns were altered by the council to redistribute flows; or if the council worked with a local fast food company to fund free coffees at peak times to take flows off the network, or if it re-programmed traffic lights to prioritize expected known individual commuter journeys? How do estimated costs of such data-driven interventions compare to physical engineering intervention costs such as building new lanes? Which routes could be replaced by new public transport links or social taxi services?

Can you make money by predicting share prices of supermarkets or other companies in the area from the number and types of journeys being made to and from them?

10.6 Further Reading

- Dean J, Ghemawat S (2004) MapReduce: simplified data processing on large clusters. In: Sixth symposium on operating system design and implementation, (OSDI)
- Gilbert S, Lynch N (2002) Brewer's conjecture and the feasibility of consistent, available, partition-tolerant web services. ACM SIGACT News 33, no 2, pp 51–59. (Proof of the CAP theorem)
- Manoochehri M (2013) Data just right: introduction to large scale data and analytics. Addison Wesley. (Excellent but low-hype guide to real big database tools)

# Professional Issues {#professional-issues .unnumbered}

Suppose that Rummidgeshire County Council is working with Rummidge University's Transport Studying Institute to understand commuter behavior on its road network, and has given the university access to its data.[1] Nigel and Nicola are data scientists who have been given access to different parts of the data at the University. Nigel is an active member of the National Rummidgeshire Party (seeking to ban immigration to Rummidgeshire) and makes weekly evening trips to meetings at its office. Nicola is a member of the Rummidgeshire National Party (which seeks Rummidgshire devolution from the UK). Their University supports devolution and immigration in its official policy, and neither staff member has declared their memberships to the University. Nigel does not agree with this policy but cannot find work anywhere else in Rummidge (because, he says, immigrants from neighbouring Borsetshire have taken all the jobs) and he needs to stay there to support his family, so he keeps his professional work separate from his beliefs.

Nigel works with RCC's ANPR data, which includes images taken on residential streets showing cars driving into their garages. One of these cameras has captured an image of Nicola undressing through her bedroom window, which Nigel secretly enjoys looking at in the evenings. Nicola works with Bluetooth data to plot congestion maps. While doing this, she notices that Nigel's phone has Bluetooth enabled in their shared office, looks up his MAC in the data, and plots his movements around Rummidgeshire, revealing his trips to the NRP.

Nicola's RNP has been collecting data from its doorstep interviews with voters, which includes reports of their address, voting intention, and estimated salary. They would like to find out where each of these people work so that they can plan a targeted workplace campaign, finding offices and factories with the most swing voters. They could do this if they has access to ANPR data. They ask Nicola to approach Nigel, and to suggest that the University might not want to promote him if they were given information about his NRP journeys, and that she may keep that discovery secret if Nigel were to let her borrow his ANPR data for an afternoon.

- What should Nicola and Nigel, and other parties do? What controls are or should be in place?

---

[1] Fictional names and events are used here, however the issues could occur in many real situations.

11.1 Morals, Ethics, and Law

Morality means making value judgments about what is good and bad. Justification of axioms for moral reasoning is notoriously problematic, with both philosophers and members of the public variously claiming inspiration from religious, evolutionary, social utility, or self-preserving assumptions. As engineers we do not typically worry about this because we work within a society that makes its own moral judgments, and encodes them into systems that result in hiring us to do paid work, within a marketplace, to deliver things that society values under these judgments. Sometimes our personal moral feelings will be involved too, for example in deciding whether to work on military transportation for one's own country or for export to highest bidders, or choosing to work on a lower-paid or *pro bona* (literally "for good" but also meaning "free of charge") project that makes us feel good. But these are personal, rather than professional, issues.

Ethics. As professionals who have already decided to work on a project, we are more concerned with professional ethics than with morality. Ethics is the study of how to practically deliver a given moral utility function, for example by translating it into national laws, professional charters, and internal company rules. Ethical reasoning is less problematic than moral reasoning, and typically proceeds by making analogies between existing understood ethics and new problems. For example: *given* that our society already considers it OK to conceal one's face in public with a niqab or motorcycle helmet, but requires police officers to be personally identifiable in order account for their actions, we might argue that *therefore* it is OK for police officers to conceal their faces on patrol as long as they display their staff number on their uniform.[2]

Law is the encoding of society's ethical judgments into enforceable rules to regulate behavior. Criminal law deals with violations against society as society, and punishes to discourage violations. Civil law deals with violations against individuals, and may enforce payment of compensation between them to restore damages. This includes damage caused by breaking legal contracts made between consenting individuals ("contract law"), as well as non-contractual civil wrongs defined by law ("tort law").

Professional codes are often voluntary rules accepted by their members, across many organizations, which provide some guarantee of the members' quality and thus increase their value and salaries. For example, chartered engineers agree to maintain honesty and integrity, and to prioritize public safety beyond strict legal requirements. Organizations may maintain additional internal codes to maintain and enhance their brands, for example the University of Leeds has a research ethics policy which staff are under contract to follow (*ris.leeds.ac.uk/homepage/2/good_practice_and_ethics*). Historically, computing and data workers have not generally professionalized under charters, but engineers have done so.

- Why do you think this is?[3]

---

[2] An example of an opposing ethical argument might be that society requires maximum quality of policing from officers paid from public funds, and high quality policing requires relationship building, which in turn requires communication though visible facial expressions.

[3] In the UK, some computer scientists are "Chartered IT Professional" members of the British Computing Society, or become Chartered Engineers, but chartership is far less popular than in other branches of Engineering. This is possibly why Data Scientists have to endure tens of hours of interviews for many jobs, while Chartered Engineers can be assumed to be of suitable quality and hired more easily. As Data Science becomes more involved with ethical as well as technical competence issues, there have been various recent suggestions for "Chartered Data Scientist" schemes.

11.2 Ethical Issues

It is likely that specific data laws are different between your countries, and that they are also in a current state of change due to political interest in Data Science. Therefore we will focus here on the general ethical principles behind them, and encourage you to think about how they appear in current laws you may be operating under, whenever and wherever you are.

11.2.1 Privacy

Portable cameras became popular in the 1880s, as in Fig. 11.1, triggering one of the first debates about data privacy. Until this technology, photography had taken place mostly inside private studios at the request of the people being photographed. The portable version enabled owners to take pictures of others in public, without necessarily obtaining their consent. This could include, for example, beachwear shots of sunbathers, which were then copied and sold for money. There was also an 1880s craze for concealed "detective cameras" which were used to take photographs of people committing legal or moral offenses, either to bring them to justice or to blackmail them. The camera owners argued that all these acts took place in public places, which by definition are open to members of the public going to see them. Opponents argued that the meaning of public space was that anyone could go there to see it, but by doing so they also identified themselves in public, enabling others to know who was watching them and to adjust their behavior accordingly.

- Do you know what the laws in your own country are on this today? Does it matter if (like guns) the camera is concealed or used openly?

As Transport Data Scientists we sometimes have access to similar images of public spaces, on larger scales than in the 1880s. We might now have CCTV and other data covering whole cities. In principle, 1880s technology could have done this given enough resources, but it is now more common. We are able to link images and sensors together to almost constantly track all individuals moving through cities and highways. Databases have also made it much faster to search though this data to observe and both particular individuals and aggregate trends. It has become cheap to store all this data for decades or perhaps forever,[4] for example to enable recovery of someone's actions as an NRP supporting teenager when they apply for a job as a RNP supporting adult, or vice versa.

- Does or should this scaling change the moral, ethical, or legal situation?
- Should data recorded from public places be made available to all the public or only to the people who record it?

---

[4]Currently, data must be kept "alive" continually by copying it between storage media which last for years or decades, and it remains an open question how to preserve it without this kind of care. Some authors have worried that our civilization might ultimately leave less data behind that paper, and even clay tablet based, historical cultures. A current research area is data storage in crystals which might last for thousands of years, see (Zhang, Jingyu, et al. *5D data storage by ultrafast laser nanostructuring in glass*, CLEO: Science and Innovations. Optical Society of America, 2013).

Fig. 11.1 Taking an early "selfie"?

11.2.2 De-anonymization ("Doxing")

Many people feel that it is OK for Data Science to work with aggregate statistics, and with individual data only if the identities of the individuals have been anonymized, or the individuals have given specific consent.

- Do you agree with this? What moral or ethical principles are behind it?

In many settings, controls such as laws or professional codes are put in place to enforce this. For example, ANPR detections on the M25 motorway are immediately hashed inside the ANPR cameras, to enable vehicles to be tracked around the network but without linking their movements to nameable car owners.

Completely unbreakable anonymisation is not generally possible. Most data recordings of individuals contains *some* information about their identity, in the formal sense of reducing the entropy of $P(individual|data)$ from the prior $P(individual)$. An extreme example occurs if we hash the licence place of the only Reliant Robin driver in Rummidge, but retain the vehicle type information which enables us to track all his movements precisely. More generally, logging vehicle type along with hashed ANPR gives us *some* information about which individual in the car, but not enough by itself to fully identify.

Information theory tells us that we can often fuse several weak sources of information into stronger information. Suppose we have several "anonymized" data sets which all contain weak information about individuals, such as their car type, commuting times, and locations visited. Using Bayesian inference, (in this case with a Naïve Bayes assumption) we can compute,

11.2 Ethical Issues

$$P(individual|D_1, D_2, \ldots, D_n) = \frac{1}{Z} P(individual) \prod_i P(D_i|individual),$$

which under Naïve Bayes, will converge to a certain identification as we fuse more and more weak data.

For most transport data collected by a single organization, such as all the traffic sensors in Rummidgeshire, such fusion is unlikely to obtain high enough accuracy for short periods of recording. It is possible however that if we include long-term historical data, such as fusing all the commutes made by an individual for 10 years, that accuracy might become obtainable in some cases.

- Exercise: estimate how many sensors, and how many years, this might be?

The situation is made worse (or better, depending on your point of view) by the existence of data aggregation companies, which now routinely buy and sell whatever data about individuals they can obtain from any sources. Depending on local data laws and individual opt-outs from them, these could technologically include true "big" data sets such as every purchase made in supermarkets, logged as reward cards or just by credit cards; every transport sensor; every action made on cookie-enabled websites including how much time spent looking at which face-identified photos, how much time looking at what on-line marketplace products (including time on each page of book previews), the text of search engine queries over decades, the natural-language parsed and topic-classified text of every email sent over decades on web-based email systems, and salaries as inferred from job titles on public "professional" social networks. If you accept legal opt-out terms and conditions on many of these services then you should assume that these companies will extract maximum value from all this data by selling it to aggregation companies – indeed they may be failing their legal duty to their shareholders if they do not extract this value. The aggregation companies will then sell it to anyone willing to buy it, which may well include Transport Data Scientists wanting to de-anonymize all their traffic data, as well as to anyone else, including advertisers, political strategists, tax enforcers, your commercial competitors, social media stalkers, foreign intelligence agencies etc. Fusing weak data at this scale may be enough to fully de-anonymize individuals from other data sets, such as traffic sensors. In this case, the scale of "big" data does make a well-defined difference to privacy – because an individual can either be identified or not, and there exists a point where the identification becomes unique or at least "beyond all reasonable doubt". Owning data is rather like owning uranium-235 – it suddenly behaves differently when large quantities are brought together.

A variation on this idea is for Data Scientists to collect their own "artisan boutique small data" as a supplement to their existing data collections, again in such as way as to provide enough information to enable de-anonymization. For example, in the M25 model of Chap. 1, we installed a few of our own unhashed ANPR cameras which enabled us to "unlock" and unmask a large amount of big data from a larger set of hashed and supposedly-anonymised ANPR data from the existing M25 motorway network. We have seem previously that this type of data might also allow new *causal* inferences to be made about the big data.

- Is use of aggregation companies or "artisan boutique small data" by government and private transport professionals moral, ethical, legal and/or professional? Should it be? Are there differences between being a buyer and a seller?

11.2.3 Predictive Analytics

A consequence of de-anonymization is the ability to make probabilistic inferences and predictions about the behavior of individuals rather than about aggregate populations. This is sometimes known as "predictive analytics". We can train generative or discriminative models on individuals historical actions, such as commuting patterns, and interpolate them into the future. As with working with demographics priors, this can be viewed as a form of prejudice or pre-judgement. In some cases, utility theory might tell us it is optimal to make intervention on information gathering actions based on these predictions. For example, models might predict that a subway passenger is about to commit an act of terrorism, suggesting optimal actions to stop and search them and they enter the network. While this may be optimal from the view of preventing terrorism, it also has consequences on civil liberties, and specifically the potential to raise demographic tensions if similar types of people are more commonly stopped than others. The book and movie, *Minority Report*, are based on this premise, in a world where people can be accurately arrested for crimes before they actually commit them. In other cases we might predict or infer that someone is – or about to become – pregnant, from changes in their movements around a city (this is already done for some farm animals moving around fields), and selling that information to aggregation companies might result in the person or their family finding out about the pregnancy before they would otherwise do. In 2013, Microsoft data scientists (Kosinski et al. 2013) reported that they were able to infer "sexual orientation, ethnicity, religious and political views, personality traits, intelligence, happiness, use of addictive substances, parental separation, age, and gender", from apparently harmless and unrelated social media posts by (consenting experimental volunteer) individuals. In 2016, Microsoft bought the entire social network database LinkedIn for $26bn.[5]

11.2.4 Social and Selfish Equilibria

Traffic data is used along with maps and GPS to route individual drivers around cities. Classical transport modelling has studied traffic (Wardrop) equilibria, usually based on the assumption that each individual is a selfish agent seeking to minimize their journey time given full knowledge of traffic on the rest of the network. Real-time traffic data presents possibilities for changing these assumptions. A data company running such a live service does not just provide data to its users, but also constantly collects data from them. If all drivers were to use such a system, then Wardrop's assumption of full knowledge would be more accurately satisfied than when drivers rely only on sight and, for example, radio traffic reports, to make inferences about the rest of the network. However most users of real-time navigation do not make their own routing decisions, but leave them to the service provider too. By outsourcing your routing to an external company, you are making the assumption that the company has your own interests at heart rather than its own. Do you know if the company is really assigning the selfish fastest route to your journey? A socially benevolent routing company might use its "God's eye view" of the network to implement a Wardrop social equilibrium, which minimizes the average journey time but at the cost of some users receiving worse-than average times, in contrast to the selfish model. Would you be happy to use a navigation program that did this, in the knowledge that everyone else was using the same one? What if other drivers had the option to switch to other systems which gave them selfish routes? The service provider needs driver data in order to understand the network, and in some cases

---

[5]For an example of a social network that cannot be mined in this way, see *www.joindisapora.com*. Diaspora is open-source and runs on a distributed network of independent servers run by trusted volunteers, so no-one is able access or sell all its data. For a search engine that (claims it) does not track or sell your search queries, see *www.duckduckgo.com*.

it will be in the company's interest to collect new data from routes where it currently has none. Would you be happy for your system to reroute you on a longer route which enables it to collect data about times along it, at the expense of your own travel time and fuel? A major social network was recently fined for deliberately filtering messages from users' friends to "make them sad", as part of an internal psychology and marketing experiment. In some cases this may have contributed to depression, suicide, or other health problems into its users, in order to improve its models and profits, with no additional consent beyond its standard sign-up small-print. It is easy to imagine fusing such ideas with transport data, for example a company might run experiments on sending drivers into all kinds of environments and routes to measure their responses. How would you feel if its system decided to do this to you when you are late for an important meeting? Have you read the small print of your contract with them to check if they can currently do this to you?

11.2.5 Monetization

How is de-anonymized data being turned into actual cash?

Low level blackmail, as in the Nigel and Nicola example, is an obvious possibility, though it is not known how common it is. Possibly, data scientists themselves should be more paranoid about it that other workers – for example if you are working or applying for jobs at a Transport Analytics team there is a high chance that other staff there will be (technically) able to pull up data about you. High profile individuals are also at risk. In 2016, The Guardian newspaper reported claims that Uber staff spying on politician's and a music star's movements, which could have been used against them or sold to newspapers or their enemies for a profit.

Differential pricing was the subject of a 2015 White House report, discussing the possibility (and possibly current practice) that personal data could be used to infer individual's net worth, buying power, and demand for goods and services. This could then be used to create a "perfect market" where every user of, say, a taxi booking system, is charged a different price. Prices would be set to extract the maximum amount of money from each customer, rather than having to use a single market rate. To a lesser extent this has happened for a long time via market segmentation. For example taxi customers with student cards may be offered a "discount" which better balances supply and demand in that segment. Differential pricing is well known in economic theory but could become a reality if the taxi company is able to buy information about your salary, workplace, job title, and commuting patterns from the aggregation companies. It is especially suited to transport service sales, because customers are already used to highly customized prices that respond to overall demand (surge pricing) and it would be difficult to separate out the differential component from this. Some people consider this a good thing, because it will extract higher fees from people who can afford them, and make lower fees available to poorer people, who would not otherwise use the resources. Other people consider it a bad thing, saying it will penalize people who choose to work hard to earn money, and reward those who do not. It may have the interesting property of creating a socialist-like redistribution but though entirely free-market means?

Information asymmetries. According to financial legend,[6] Nathan Rothschild was able to make a large profit in 1815 by relaying news of the result of the Battle of Waterloo to his trading desk via carrier pigeon, and buying post-war stocks before his competitors read about it in the newspapers. In modern finance, the same idea is used by "high-frequency traders" who try to buy, communicate and analyze data before everyone else. Classical economics is based on the assumptions that everyone in

---

[6]Though not necessarily to reality, see B. Cathcart, *The News from Waterloo: The Race to Tell Britain of Wellington's Victory*, Faber & Faber 2015.

the market has access to, and can instantly compute optimal actions from, the same data at the same time, which are not true. Traders can now use transport data such as vehicle counts or ANPR detections to infer who is shopping in stores before they release their official accounts to the public. This enables them to predict these accounts in advance, and transfer money from your pension fund to their bonuses by buying early then selling on to your pension fund at a higher price. They argue that this is a good thing, because it serves society by making the market prices more accurate at all times and not just on the days that reports are published. Society pays them for fusing the new information, which they have worked hard to collect, into the market price for others to see. Information asymmetry is a famous factor in used car trading, where the previous owner knows much more about a particular vehicle's state (whether it is a good "peach" or a wrecked "lemon") and hence its value than the buyer. This loss of information is why new cars lose around 20% of their value when they are driven off the forecourt. It is possible that new bulk data from telematics and roadside sensors might be useful to re-balance this in the future and make the used car market more efficient.

Insurance. The car insurance industry collected 671bn USD of premiums in 2014 (Finaccord report 2015) and makes its money by predicting who is likely to make claims. Traditionally its predictions have been based on simple demographic information plus historical claims data. But they may become much more accurate if they are based on bulk personal data bought from aggregation companies. EU insurance companies are banned by law from discrimination against people on some properties such as gender. But it is not clear how this can be enforced if bulk data is used which, as in Microsoft's study, could be used to infer such properties indirectly. The life insurance industry will also be interested in inferences about your personal life which could predict risk of early death, including public transport logs of trips to fast food restaurants, doctors, chemists, and different types of nightclubs. The ethics of predictive analytics are important here – as in policing, an insurer might infer that you are at risk of drink driving from your Saturday evening telematics trips to a city center, even if you are in fact a medical shift worker at those times (Fig. 11.2).

- Which of the above are moral, ethical, legal and/or professional?

Fig. 11.2 From *www.pinterest.com/pin/1829656072184053*

"Your recent Amazon purchases, Tweet score and location history makes you 23.5% welcome here."

11.2.6 Ontological Bias

Design of ontologies and ETL processing are not value-neutral, and the people who do them will enforce a particular way of conceptualizing the world onto analysts using their databases. For example, transport database administrators in the UK are probably not interested in the gender or religion of drivers imaged by their ANPR cameras, and will discard this information when they reduce the detection from pixels into licence plate characters. But in Saudi Arabia it is a crime for women and non-Muslims to drive in some areas (Fig. 11.3), so this data is important and would be useful to extract and store.

It is possible that the personal attributes of data managers can consciously or subconsciously influence their choices, with men and women, immigrants and natives, religious and non-religious people, geeks and non-geeks all considering different feature to be salient. For example the advice given in this newspaper column of Fig. 11.4 seems quite optimal given a particular decision about what data to extract and act on.

This may cause problems if an organization's data managers are drawn from a non-diverse pool of people, especially if it is different from the people who will work on the analysis.

Related to ontology bias is a problem of believing in formal data too strongly. Especially in classical database design, with its formal administrative processes and high-status data managers, there can be a tendency for data analysts to believe whatever is in the database as a completely true and consistent model of the world. They forget that both the choice of ontology, methods of data collection, and interpretation of the meanings of the properties are all very human constructions. For example a property like "number of cars" in a database table might refer to the total number counted in a minute, or day, or month; or to the number of *types* of cars seen in a day or produced by a manufacturer, or to any of these including trucks and motorbikes as "cars", or excluding small vans or SUVs. You have probably worked with other people's data and very quickly hit problems like this. You can ask the data provider to define the terms in English, but English is also a human construction which can be just as vague, or ever vaguer, than SQL. For example in March 2017, a USA company faces a millions of dollar court case over their definition of what work is counted as overtime, due to the possible absence of a single comma after "packing" in,

"The canning, processing, preserving, freezing, drying, marketing, storing, packing for shipment or distribution of: (1) Agricultural produce; (2) Meat and fish products; and (3) Perishable foods."

Fig. 11.3 Religious transport ontology in Saudi Arabia

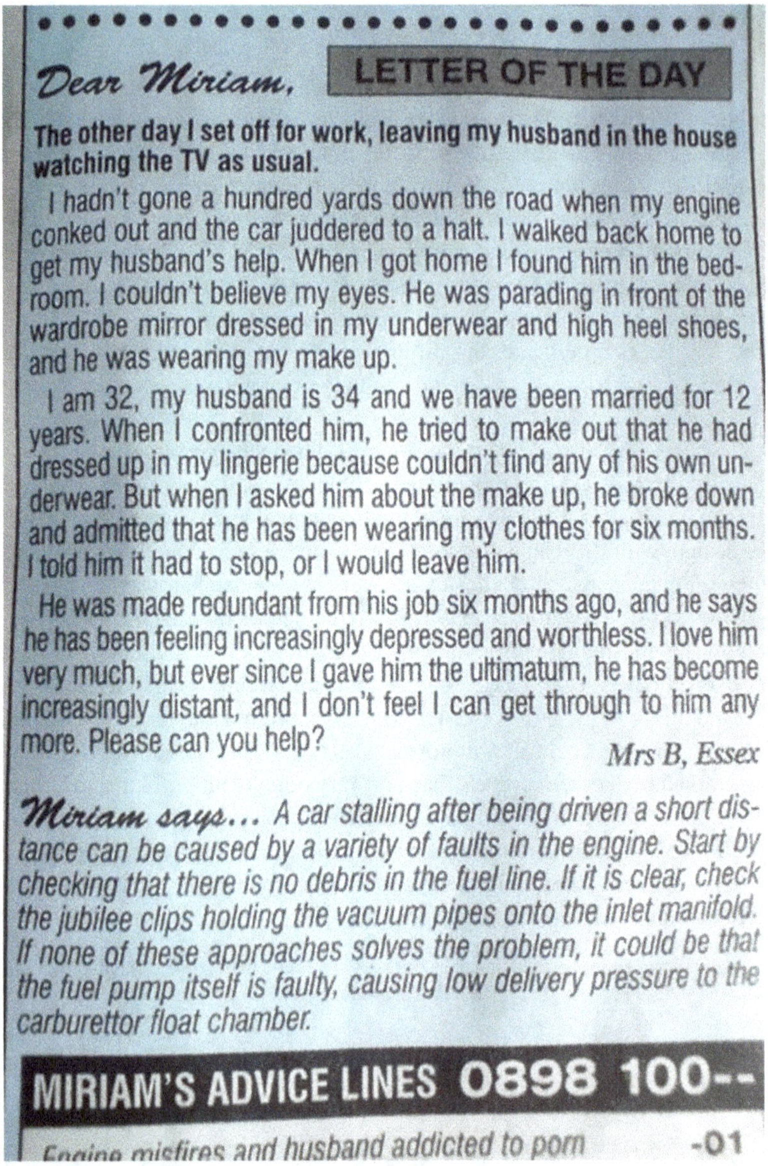

Dear Miriam, LETTER OF THE DAY

The other day I set off for work, leaving my husband in the house watching the TV as usual.

I hadn't gone a hundred yards down the road when my engine conked out and the car juddered to a halt. I walked back home to get my husband's help. When I got home I found him in the bedroom. I couldn't believe my eyes. He was parading in front of the wardrobe mirror dressed in my underwear and high heel shoes, and he was wearing my make up.

I am 32, my husband is 34 and we have been married for 12 years. When I confronted him, he tried to make out that he had dressed up in my lingerie because couldn't find any of his own underwear. But when I asked him about the make up, he broke down and admitted that he has been wearing my clothes for six months. I told him it had to stop, or I would leave him.

He was made redundant from his job six months ago, and he says he has been feeling increasingly depressed and worthless. I love him very much, but ever since I gave him the ultimatum, he has become increasingly distant, and I don't feel I can get through to him any more. Please can you help?

Mrs B, Essex

Miriam says... *A car stalling after being driven a short distance can be caused by a variety of faults in the engine. Start by checking that there is no debris in the fuel line. If it is clear, check the jubilee clips holding the vacuum pipes onto the inlet manifold. If none of these approaches solves the problem, it could be that the fuel pump itself is faulty, causing low delivery pressure to the carburettor float chamber.*

MIRIAM'S ADVICE LINES 0898 100--

Engine misfires and husband addicted to porn -01

Fig. 11.4 Don't let transport engineers extract the salient data for analysis tasks outside their domain

It is possible that these issues will be reduced if Data Science moves away from the formal Codd model to the "data lake" model of storing raw, unprocessed data and having analysts do their own ETL on the fly. However this creates more work, and often duplication of work, for analysts, which the Codd model was originally designed to reduce.

11.2.7 *p*-hacking

If a data scientist proposes a model M_1 with a free parameter θ, is has a better chance of fitting the data than a model without the free parameter, M_0. This is because the parameter can be adjusted to one of many values to obtain the best out of many possible fits, while M_0 only gets one chance. Similarly, a model with n parameters, M_n, has even better chances. Effectively, M_1 and M_n represent whole collections of models, with different parameter instantiations. Bayesian theory includes an automatic and natural "Occam's Razor" penalty (known as an Occam Factor or Bayes Factor) as we expand the model posterior to include the parameter,

$$P(M|D) = \frac{P(D|M)P(M)}{P(D)} = \frac{\int_\theta d\theta\, P(D|M,\theta) P(\theta|M) P(M)}{P(D)}.$$

As the possible parameter space gets larger, the prior of it taking any particular value $P(\theta|M)$ gets lower, which is traded off against the better fits of $P(D|M, \theta)$.

This is good science. However due to external management factors – notably the pressure to publish significant findings in academic research, or for commercial analysts' bonuses or prestige to be tied to significant discoveries – data scientists are often strongly motivated to "game" these probabilities by a process known as "p-hacking". In p-hacking, the data scientist pretends that they had originally proposed a single, non-parametric model, $M_{\theta=\hat{\theta}}$, where the optimal value of θ from M_θ is simply assumed and built into the model, rather than appearing as a free parameter. i.e. they assume that,

$$P(\theta|M) = \delta(\theta = \hat{\theta}).$$

This then allows them to report a massively higher $P(M|D)$ and claim that their model is better than other candidates. The choice of $\hat{\theta}$ can take place via various processes with various levels of dishonesty. Most blatantly, the data scientist can simply train their parametric model M_θ on the data, then propose a new model $M_{\theta=\hat{\theta}}$ afterwards, inventing some story about why they assumed $\hat{\theta}$ for a reason other than their training. This is simple fraud. More subtly, the data scientist might try out hundreds or thousands of different non-parametric models over a year, choose the one that performs best on the data, and report it without mentioning the number of alternatives tried out. Even more subtle is the "file drawer effect" where a community of hundreds or thousands of data scientists each propose one such model; most of them are terrible but one gets lucky as is submitted to and published in a journal, while the rest are not submitted or heard about, and are kept in the data scientist's "bottom drawers" instead. This has been proposed as an explanation for why *half* of all published Psychology results are apparently unreproducible and possibly false.[7]

One reason why Data Science can be an unpleasant profession is that we can never be completely sure that our software and findings are correct. It is always possible that your apparent result has occurred due to a bug in your code or a glitch in the data, rather than to the exciting new insight which it suggests. Formal verification of computer programs has been an active research area in Computer Science for many decades but has still not reached a state which would enable it to guarantee that a Data Science analysis program really works. Typical software is estimated to have roughly one bug per 100 lines. For example, your finding that driving safety correlates with mobile phone use might just be due to a bug which mislabels some other variable name as "mobile phone use". (The best way to find these bugs is to work closely with a business domain expert and have them explore the findings, pose new questions, and make assertions which can be checked against the same data and results for inconsistencies. Confidence is gained in the system as more of these checks appear to be consistent, though it is an expensive process and can never reach complete certainty.) Sometimes when a bug may have led to an interesting (publishable, and/or career-advancing) result, it is extremely tempting to "turn a blind eye" to any minor strange behavior of the program. In contrast, if the analyst does

---

[7]Open Science Collaboration, *Estimating the reproducibility of psychological science*, Science, 2015. DOI: 10.1126/science.aac4716.

not like the apparent result, there is more incentive to go looking for bugs in more detail, and keep fixing them until a more likable result is found. Effectively, the collection of versions of the program in various states of debugging may function as a set of models in another form of *p*-hacking.

p-hacking is related to model selection in Philosophy of Science. In *The Structure of Scientific Revolutions*, the philosopher Kuhn described the process of model failure and replacement in science. Initially, a model may fit the known data quite well. Over time, new data are found which fit less well. At this point, some scientists will want to modify the model, for example adding new parameters or changing old ones, to better fit the new data. Other scientists will propose new models with fewer parameters that better fit the data. Politics will occur as the community decides which data is *more important to explain* with the models. The scientists proposing new parameters are accused, in Data Science terms, of *p*-hacking, by the new model proponents.[8]

p-hacking ranges from outright fraud to subtle social causes. It is best controlled, in ideal cases, though use of an independent party's custodianship of an independent test set of data that is not made available to the proposing data scientist, and is only ever consulted once, and *after* the proposer's claimed model has been stated in public. This ideal setup is possible in Data Science competitions such as those administered by the company Kaggle, who act as a trusted custodian in this way. Another way to achieve it is for data scientists to publish validation studies of other data scientist's previously claimed and published models. But in other cases it might not be possible to hold back public data or to collect new data, and some element of trust and professional ethics is required from data scientists. It is possible that future "Chartered Data Scientists" will sign up to such standards, and educate the public to be wary of possible *p*-hacking by non-chartered authors.

11.2.8 Code Quality

Data Science code is often written in a very different style from regular Software Engineering. In Software Engineering, great care is taken to make the code not only correct but also to make it robust to possible future changes and extensions. For example, instead of working with a Car class, it might build in a Vehicle class and make the Car a special case (subclass) in order to allow other types of vehicle to be added in the future. It might also build in various types of automated error checks and recoveries to allow the code to run or recover for possible future inputs that do not occur in the present task's input. It takes care to use human-readable comments and variable names, and to write code in a readable way. All of these future-proofing styles require more human time and effort to produce than it would take to write a single-use script to deliver only the currently specified task on the currently available data. Data scientists are thus faced with a professional decision about which style to use. Single-use scripts will reduce the immediate cost to their organization (especially in cases of "could you get me those numbers by this afternoon?"); but if the organization comes back days or years later and asks for some new variant of the task, then a Software Engineered system will be quicker and cheaper to alter for its new requirements. Many programmers consider that building in good Software Engineering style is a matter of personal or professional pride, regardless of the need for future-proofing. Some managers get annoyed, if they ask for a quick calculation of something that could be done in an afternoon, and they get a heavyweight engineered system that takes weeks, at a day-rate, to deliver. However these same managers then get annoyed again the following year when then system isn't able to compute a new, variant, result quickly and easily. Data scientists need to make a judgment about what approach to take in these situations, and it is possible that future charters will help them here. Perhaps chartered

---

[8]This is currently the situation in String Theory, which is accused by its detractors of effectively *p*-hacking from a space of 10^{500} parameters, as in L. Smolin (2007), *The Trouble with Physics*, Mariner Books.

data scientists will be assumed, and costed, to always build slower but higher quality code rather than single-use scripts, in order to enhance the quality of the profession as a whole though at the expense of delivering small, fast results, which would then be delegated to "business analysts" or other, cheaper, "para-data-science" staff.[9]

11.2.9 Agency Conflicts

Government IT systems form the basis for many Transport Data Science applications, and their creation is notoriously expensive and error-prone. Failed government IT systems have led to major political embarrassments. Commissioning an IT system is usually a case of hiring a "professional", in the sense that the hirer is a manager with little or no technical skill, and is ultimately dependent on trust in the skilled hire. Typically "the professional" is one or more large contracting or consulting firms. The history of government IT systems is littered with stories of these firms blaming one another for failures (Goldfinch 2007), arguing over who is responsible through layers of subcontractors and sequences of replacement leaders. Creating detailed "specification" documents to describe these responsibilities creates more work for all concerned, and is itself prone to errors – if anyone really understood the specification in perfect detail, they could just write it down directly as executable code rather than English in the first place. In some cases the level of "professionalism" seems no higher than found in cut-throat domestic tradesmen who blame each other's "shoddy work" to justify their own large fees, especially when coupled with issues of "research code" quality as above. This is a notably different attitude to those maintained in other, more established, professions such as Law and Medicine, where even members of directly competing firms rarely criticize each other's work.

11.2.10 Server Jurisdiction

Data analysis is moving increasingly to true big data systems – running on large distributed compute clusters – which are often provided as "infrastructure as a service (IaaS)" (or "cloud computing") by a specialist external company. Often this company will be based in a different country from the organization doing the analysis work. Further, the physical machines hosting the data could be in one or more additional countries. The analyst might be working from home or from a hotel in yet another country, and the data being analyzed might be from several further countries. Uniquely to big data analysis, different parts of the computation itself could be performed and fused together in different countries. For example, and insurance company might want to move all its gender-specific algorithms out of the EU but have them report some proxy variable back to the EU which effectively contains the same information. In such cases it might not be clear who is or should be responsible for what aspects of data storage and use. The different countries might all have different moral, ethical and legal systems – which ones should the analyst operate under?

11.2.11 Security Services

Transport data may be of particular interest to security services, for example many terrorist incidents target public transport, and data from both public and private transport may be useful to track a

---

[9]There is a famous internet parable called "The Computer Scientist and the breakfast food cooker" about this dilemma, currently mirrored at *http://philip.greenspun.com/humor/eecs-difference-explained* and many other sites.

suspect's movements. Security services typically operate under specific laws which allow them to access otherwise restricted data in return for safeguarding it and guaranteeing it will not be used for other purposes (such as enforcement of minor offenses like speeding). Some people, such as Edward Snowden, have felt that security services have acted unethically, incompetently, or illegally in their handling of personal data. Others point out that despite large public leaks about these systems in recent years, there has been no evidence – for the UK's services at least – of acting unethically, incompetently, or illegally, as they deliver work that their citizens pay them to do to keep them safe, and thus that the leaks have actually strengthened their ethical and technical reputations.

11.3 UK Legal Framework

Disclaimer: This section gives a short overview of relevant legal concepts in the UK. It does not constitute legal advice. You should take legal advice when you are faced with real problems. You should consult your organization's Data Controller in the first instance before beginning any study if you are not sure of the law.

11.3.1 Data Protection Act 1988

Under the UK Data Protection Act 1988, everyone responsible for using personal data in the UK has to follow strict rules called "data protection principles". They must make sure the information is:

- used fairly and lawfully used for limited, specifically stated purposes
- used in a way that is adequate, relevant and not excessive
- accurate
- kept for no longer than is absolutely necessary
- handled according to people's data protection rights
- kept safe and secure
- not transferred outside the European Economic Area without adequate protection

Personal data may only be processed if the subject has given consent. "Personal data" means data which relate to a living individual who can be identified:

(a) from those data, or

(b) from those data and other information which is in the possession of, or is likely to come into the possession of, the data controller.

The data controller is a named individual who takes responsibility for a data set. Often this role is centralized by an organization, assigning one individual to take the legal responsibility for all data work done across the organization.

The act does not, however, define "consent". It is usually taken to mean written permission, though it has been suggested that, for example, placing signs in public places such as "by being in this space you consent to being recorded" may be appropriate. This is important for transport data in particular. Websites typically place details of what is being consenting to in massive legal texts which most people just click "I agree" to, without ever reading them.[10]

---

[10] Attention was brought to the consent terms and conditions for the iTunes program by turning them into a 108 page comic book, in R. Sikoryak, *Terms and Conditions*, Drawn and Quarterly, 2017. In 2014, several London commuters consented to transfer ownership of their first-born children to a "free" wifi hotspot provider by clicking "I agree", *metro.co.uk/2014/09/30/people-would-offer-up-first-born-child-for-free-wifi-stunt-reveals-4887827*.

11.3 UK Legal Framework

The act gives subjects of personal data the rights:

- to be informed if data about them is being processed
- to be informed about the purpose and users of the processing
- to find out what data is owned about them and receive a copy[11]

The act does not apply to non-personal data.

There are exceptions for:

- "research purposes" includes statistical or historical purposes, under all of the conditions:

  1. that the data are not processed to support measures or decisions with respect to particular individuals,
  2. that the data are not processed in such a way that substantial damage or substantial distress is, or is likely to be, caused to any data subject,
  3. the results of the research or any resulting statistics are not made available in a form which identifies data subjects or any of them.

- crime monitoring and prevention, e.g. CCTV data used solely for this purpose.
- journalistic, literary and artistic purposes – including use of photography for these purposes, though not for general use.
- other cases including security, legal, taxation and heath uses.[12]

CCTV used for non-exempt purposes in public places should be used under notification of the Information Commissioner's Office, including of its intended purpose and duration. For example, you may not re-use CCTV data intended for aggregate traffic statistical "research purposes" to track which of your staff are visiting political party premises after work. There is no distinction between hidden and visible cameras. Private landowners may impose whatever additional conditions they wish on data collection including images, taken within their property. Human Rights law gives an "expectation of privacy" which may be violated by collecting data about a private place from outside of it.[13]

There is stronger legal protection for "sensitive personal data", such as ethnic background, political opinions, religious beliefs, health, sexual health and criminal records.

The definition of (b) above is extremely problematic given the previous discussion of big data aggregation, and is the subject of recent legal research and debate. The notion of (1) is also problematic as transport models move towards per-person predictive analytics.

11.3.2 General Data Protection Regulation (GPDR)

At the time of writing, the European Union's new General Data Protection Regulation (Regulation (EU) 2016/679) has been adopted by member states including the UK and is due to come into force in

---

[11] With exceptions for some legal, security, and tax data.

[12] There may be other laws which can overrule Data Protection too. For example, under Human Rights law, disabled persons have certain rights to enable them to work and communicate, which may require their trusted agents to access data about them.

[13] For detailed recommendations for CCTV compliance with the Data Protection Act, see the UK Home Office 2013 *Surveillance Camera Code of Practice*, and the Protection of Freedoms Act 2012.

2018, and aims to address some of the above problems with UK and other member states' laws. Unlike EU Directives, a Regulation is immediately binding on all member states when it comes into force.

It applies to personal data which is collected or processed in the EU and to personal data about EU citizens held anywhere in the world. It defines "personal data" as "any information relating to an individual, whether it relates to his or her private, professional or public life. It can be anything from a name, a home address, a photo, an email address, bank details, posts on social networking websites, medical information, or a computer's IP address". It has exemptions for some law enforcement uses.

Consent by subjects for collection and processing of data about them is defined more strongly than in the Data Protection Act, and requires a positive "opting-in" act of consent rather than passive acceptance of a notice.

It provides a "right to be forgotten" which allows subjects of data to force it to be fully deleted, and a right to "data portability" which means that subjects can demand data about them to be provided in an open format which can be transferred to their own, or to competitors' systems. New rights are provided to subjects of data to be given explanations of and to challenge actionable decisions which have been made about them by automated systems such as inferential and discriminative models. There is currently still debate about what this means in practice for "black box" discriminative methods, which is likely to lead to interesting test cases in the near future.

Data Protection Officers (similar to Data Controllers in the UK DPA's definition) have new responsibilities to design data systems to protect data by default, and to notify subjects in the event of data breaches. Pseudonymization and encryption of personal data is strongly encouraged. Fines for non-compliance with the GDPR can be up to 20 million Euros.

11.4 Role of the Data Scientist

Supposedly, "Data Scientist" is the "sexiest profession of the 21st century", but who exactly are Data Scientists?

We looked at a list of related roles at the start of this book, and have avoided defining "Data Science" other than as a fuzzy cluster of skills and interests, including statistics, machine learning, database design, database administration (DBA), the computer science of parallel, distributed big data systems, data visualization, and business consulting.

Historically, these roles have been performed by different specialists. Mike the manager asks Anna the analyst to report on where to build new traffic lights; Anna turns this into statistical questions and SQL queries, and works with Dave the DBA who builds the ontology and makes data available to her within it. The Codd model of databases has enforced this, with a "modernist" culture that databases should be very clean, logical, consistent, and only editable by trained and powerful DBA professionals. This has arguably led to problems when analysts wish to conceive of data in different ways than the DBAs.

The big data movement has emphasized a "postmodern", do-it-yourself approach to data, where analysts work directly with masses of unprocessed, raw, files, performing their own interpretation of its structure on-the-fly along with their analysis. This has required analysts to understand more about the underlying computer science and software systems of databases, as well as their more traditional roles of writing SQL queries and doing statistics on the results.

The very notion of doing "statistics" is also being replaced by doing Bayesian "inference". Powerful computers and Bayesian libraries such as PyMC3 allow a much wider range of models and questions to be tested by relatively non-specialist analysts, who can now easily design graphical Bayesian network models, and ask the computer to give probabilities of whatever they are interested in, rather than developing new specialist statistics theories and estimators about them. In particular this has enabled

11.4 Role of the Data Scientist

Fig. 11.5 From *www.dilbert.com*

data professionals from more technical backgrounds, such as DBAs, to become more involved with analysis.

A recent research area is "NewSQL", which seeks a "reconstructionist" setup that looks just like the Codd model to the analyst, but transforms it automatically into real big data parallel, distributed systems under the hood. HiveQL works towards this goal, though the computer science is very hard. If this happens, then we may see a return to more clearly defined specialist roles, with statisticians going back to being statisticians and DBAs going back to begin DBAs. Or perhaps one day databases and inference systems will become simple enough, and managers will become smart enough, for managers to pose and receive answers to data questions without the need for technical intermediaries (Fig. 11.5).

11.5 Exercises

- What are the moral, ethical, legal and professional issues involved with the analysis of Bluetooth data for a county council? How can or should they be resolved?

11.6 Further Reading

- Goldfinch S (2007) Pessimism, computer failure, and information systems development in the public sector. Public Adm Rev 67.5:917–929
- d. boyd [sic], Crawford K Six provocations for big data
- USA Presidential Office report (2015) Big data, differential pricing, and advertising. https://obamawhitehouse.archives.gov/sites/default/files/whitehouse_files/docs/Big_Data_Report_Nonembargo_v2.pdf
- Shapiro C, Varian HR (1998) Information rules: a strategic guide to the network economy. (A modern classic on the economics of information and competitive data strategies)
- Kosinski M, Stillwell D, Graepel T (2013) Private traits and attributes are predictable from digital records of human behavior. Proc Natl Acad Sci 110.15:5802–5805
- Uber employees "spied on ex-partners, politicians and Beyoncé". The Guardian, 13 Dec 2016

Index

A
Abduction, 116
Airline pricing, 4
ANPR, 1, 6
Antisocial Unpleasant Driver Indicator, 139
Application Programmer Interface, 28
Aristotle, 31
Automated Number Plate Recognition, 96
Autonomous vehicles, 139

B
B+ tree, 52
Back-propagation, 102
Bayes' theorem, 78
Bayesian theory, 75
Bayesian Network, 80
Big data, 1, 7, 147
Black box, 93
Bluetooth, 40
Boutique artisan small data, 147

C
CAP theorem, 151
Car insurance, 84
Causality, 86
Chomsky hierarchy, 44
Class, 31
Cloud, 106
Cloud computing, 154, 177
Coastline paradox, 118
Codd, 37
Colours, 125
Conditionals, 20
Context-Free languages, 46
Context-Sensitive Grammars, 47
Correlation, 6
CSV file, 27, 50
CUDA, 103
Cynical views, 10

D
Data frames, 23
Data lakes, 152
Data marts, 150
Data munging, 43
Data preparation, 43
Data Protection Act, 178
Data Science, 1, 7
Database, 27
Dates and times, 48
Debugging, 24
Decision Trees, 97
Deduction, 116
Deep learning, 103
Differential GPS (DGPS), 60
Differential pricing, 171
Discriminative classification, 93
Distal cause, 87
Distributed relational databases, 157
Docker, 11
Doxing, 168
Dredging, 104

E
Ensembles, 104
Enterprise data, 147
Entropy, 98
Ethics, 10, 166
Experiment, 87
Extract, Transform, Load, 43

F
Facts, 33
Fake science, 84

File structures, 28
Files, 20
Finite State Machines, 44
Floating point numbers, 48
FPGA, 103
Functions, 21

G
Gamma distributions, 77
Gaussian distribution, 94
Gaussian Processes, 113
General Data Protection Regulation, 179
Geodesy, 57
Geographic Information Systems (GIS), 61
Geography, 107
GeoJSON, 65
GeoPandas, 66, 67
Global Navigation Satellite Systems (GNSS), 59
GPS, 59
GPU, 103
GPy, 119
Graphs, 23
Grid computing, 153

H
Hadoop, 158
HDFS, 155
Head Up Displays, 139
Hello world, 18
Hit-and-run incident, 78
HLIPR, 99
HTML, 50

I
Indices, 147
Induction, 116
Inference, 75
Information asymmetries, 171
Injection attacks, 47
Insurance, 172
Integrated Development Environment, 12, 18
IPython, 17
Itsleeds, 11

J
JSON, 50

K
Kernel, 96
Kriging, 113

L
Law, 166
Leaflet, 141

Legal framework, 178
Libraries, 21
Lidar, 4
Linear Discriminant Analysis, 95
Loops, 20

M
M25 motorway, 1
Machine learning, 93
Managers, 181
Map-reduce, 154
Maps, 131
Markov Chain Monte Carlo, 80
Markov Random Fields, 109
Matplotlib, 23
MCMC, 80
Medium-sized data, 147
Millennium Bug, 43
Model comparison, 89
Modernism, 34
Modules, 21
Morality, 166
Motorway journey times, 76

N
Naïve Bayes, 96
National Grid, 58
National Marine Electronics Association (NMEA), 50
Nearest Neighbor, 95
Neural networks, 99
Normalization, 36
NoSQL, 156

O
Object-oriented programming, 21, 33
Occam Factor, 175
OLAP, 150
Ontology, 29
Open Geo-spatial Consortium, 63
OpenCL, 103
OpenStreetMap, 65
Origin-Destination analysis, 1

P
Pandas, 23
Partitioning, 148
Pgrouting, 66
P-hacking, 104, 174
Philosophy, 29
Plotting, 23
Popper, 90
Posteriors, 79
PostGIS, 66
PostgreSQL, 37
Post-modern, 34, 152
Potholes, 4

Index   185

Predictive analytics, 170
Priming, 116
Priors, 79
Privacy, 167
Prosecutor's fallacy, 79
Proximal cause, 87
Psycopg2, 40
Pushdown Automaton, 46
Python, 15

R
R, 17
RAID, 149
Real scientist, 88
Real Time Localization Systems (RTLS), 60
Real-Time Kinematic (RTK), 60
Recursively Enumerable, 47
Regular languages, 45
Relational model, 28
Replication, 149
Road from Athens to Thebes, 29
Routing, 114
R-tree, 63
Rummidgeshire County Council, 165

S
Saunderson, Nicholas, 78
Scene analysis, 103
Self-driving cars, 139
Sense and reference, 36
Sharding, 151
Simple features, 63
Simpson's paradox, 118
Slippy maps, 136
Small data, 147
Snowden, Edward, 178
Spark, 161
Spatial index, 63
Spatial statistics, 108
Spyder, 18
SQL, 37

Strings, 48
Structured Query Language, 37
Sun Grid Engine, 153

T
Tab character, 20
Taxi, 5
Template matching, 96
Things, 29
Tobler's First Law of Geography, 107
Traffic flow maps, 133
Traffic lights, 90
Transport accidents, 15
Turing machines, 47
Type, 31

U
Unified Modeling Language, 35
Universal Transverse Mercator, 58
Unsupervised methods, 105

V
Vehicle emissions, 93
Virtual desktop, 11
Visualisation, 125

W
Wardrop equilibrium, 170
Watson, 51
Web 3.0, 50
Well-Known Text (WKT), 65
WGS84, 57
Wittgenstein, 33
World Geodetic System, 57

X
XML, 50

GPSR Compliance

The European Union's (EU) General Product Safety Regulation (GPSR) is a set of rules that requires consumer products to be safe and our obligations to ensure this.

If you have any concerns about our products, you can contact us on

ProductSafety@springernature.com

In case Publisher is established outside the EU, the EU authorized representative is:

Springer Nature Customer Service Center GmbH
Europaplatz 3
69115 Heidelberg, Germany